The Voice
Of The Crystal

How To Build
Working Radio Receiver Components
Entirely From Scratch

Written and Illustrated By
H. Peter Friedrichs

This book is dedicated to Lynn, for all her patience.

Copyright (c) 1999 H. Peter Friedrichs. All Rights Reserved

ISBN Number 0-9671905-0-9

No portion of this book may be reproduced through any means
without the written permission of the author.

This book is a compendium of practical knowledge associated with
the construction of vintage radio equipment. It is presented for
academic purposes only. Radio construction requires tools,
materials, and procedures that may be hazardous if attempted by
unqualified individuals. Not all of the information contained here has
been verified, and none is endorsed. The reader assumes full
responsibility for his/her safety. Any trademarks mentioned are the
property of their respective holders

Table Of Contents

Chapter 1
An Introduction

Welcome! Crystal radio is a "lost" (abandoned!) technology that allows the experimenter to decode and listen to radio broadcasts with extremely simple, easily-built equipment. Crystal radios use no active components, so no tubes, transistors, or batteries are required.

At the dawn of radio, many such receivers were built by thousands of people. In recent times, hobbyists have rediscovered their charm, and the useful lessons these old-timers have to teach. Interest in crystal radios is blossoming again.

There are now technical societies dedicated to the crystal radio, where members can exchange information and experience. There are numerous books available on the subject as well. Many are recent, though some of the very best are reprints of books originally published in the 1920's. These volumes are filled with valuable and interesting technical information. Often, they feature instructions so specific, that if followed, almost *guarantee* a working radio in the end.

Why bother tinkering with century-old technology? If you ask a dozen builders of crystal radio sets, you will undoubtedly receive a dozen different answers. For what it's worth, I have a few of my own.

First off, there is something wonderful, if not mystical, about the notion of snatching electromagnetic waves from the air and listening to them without the aid of tubes, transistors, or integrated circuits. In truth, the equipment necessary to accomplish this feat of magic can be of the most humble origin. At the end of every driveway, on garbage day, sits the material for dozens of radio receivers. All you need is some patience and the necessary know-how.

Figure 1-1
A genuine old timer

Second, there is a great deal of nostalgia associated with crystal radio technology. Playing with old-style radio sets is like peeking through a window into another era. In the earliest days of radio, the world was a different place. People thought differently, acted differently, and dressed differently. Life was different. Values were different.

Crystal radios represent an infant technology from more innocent time. At the very least, genuine equipment from this era shows workmanship and attention to detail that is completely lost in this age of mass-production and molded plastics. For many of us, there is joy in rediscovering that craftsmanship, investing such attention in the radios that we build.

There is a third reason, and to me it's the most important of all. The point is a bit deep, so let me present it this way:

Some time in the late 1950's, a fellow by the name of F. B. Lee threw a switch and fired up his linear accelerator, more commonly known as an "atom smasher". Driven by a 350,000 volt power supply, the beam was more powerful than that used by the famous physicists

Cockroft and Walton to transmute elements. Where did this monumental event take place? A government research facility? Some mega-corporate laboratory? M.I.T. or Berkeley? No, the machine was home-built from glass tubing, copper pipe, gallon cider jugs and an old refrigerator compressor.

What does a home-built atom smasher have to do with crystal radios? As it turns out, a great deal. The curiosity, ingenuity, and inventiveness that led Lee to undertake his project was typical of the average radio experimenter in the early days of radio. In those days, parts were expensive, if not unavailable, so experimenters frequently designed and home-brewed many components themselves. The field was wide open, there was vast uncharted territory, and almost any radio hookup exhibited advantages or disadvantages that contributed to a growing body of knowledge as a whole.

It seems that for most of the 20th century, weekend scientists and midnight engineers have tinkered, designed, and experimented in attics, cellars, and garages across America. The pioneering, do-it-yourself, try-it-and-see attitude used to be prominent characteristic of the American psyche.

If you ever have the opportunity, try to borrow a copy of The *Amateur Scientist*, published by Simon and Schuster in 1960. Lee's project, and several dozen others fill nearly 600 pages. The diversity and sophistication of the scientific equipment constructed in America's homes would astound you --- and all in an age before the microchip and the desktop computer.

I can't quite place my finger on it, but I sense that something has changed in the last few decades. The number of people engaged in this type of hobby experimentation seems to have dwindled. Witness the demise of Heathkit, for example. It astounds me that in all of America there wasn't enough interest to keep well made electronic kits a viable business.

At no time in the Earth's history has there been a society that is more technically advanced than ours. Our populace, as a whole, is probably more educated than at any point in our past. Just the same, across America, there are millions of VCR's whose displays blink "12:00" because the owners can't figure out how to set the clock. I submit that never has the average American been more clueless about the technology that makes his quality of life possible. There is no shortage of information, we just don't know how to *think* anymore.

I have noticed similar shortcomings in some of the engineers I have worked with. When faced with problems for which there are no off-the-shelf solutions, they are stymied. They are educated, and knowledgeable, but they don't know how to think. So, they implement cumbersome and expensive solutions instead of creative, simple, and cost effective ones.

The art of thinking is like the art of music or the art of dance. It can only be refined through *practice*. The more you think, the better you think. The challenges associated with crystal radio construction exist on so many different levels that there is a place for *anyone* with a fundamental curiosity, whether a graduate engineer or a casual hobbyist.

Many aspects of radio propagation can only be described with advanced calculus. At the same time, even a small child can wind a crude coil on an oatmeal box, and receive intelligible signals. Every discovery leads to two new ideas and two more questions, which can be pursued as far as your time, money, and patience will allow.

So, I arrive at the conclusion of my argument and my final point: Yes, crystal radio is old news, and may be of limited value as an educational experience. However, it excels as an exercise in creative and logical *thought*. The ability to think creatively has application *everywhere*.

The purpose of this book is to share with you some of my crystal radio adventures. It is not a cookbook, with detailed instructions and dimensioned blueprints. There are plenty of other books that will provide that for you. In this book, my goal is to illustrate the application of the creative thought process to basic groundwork theory. Combined with the raw materials found in every garage or garbage can, the result is some interesting working equipment, and a new look at problem-solving. Few lessons in life are this much fun!

You have not read a book like this before.

Chapter 2
The Challenge

B esides being an electrical engineer by profession, I have been an incurable experimenter and gadgeteer since childhood. I love old science and technology books, and have a special fascination for people, past and present, who can build something out of nothing. In fact, I have something of a reputation for this myself.

Not too long ago, I was inspired by some old books to mess around with crystal sets once more. This time, however, my work would have a special twist. I decided that it would provide an interesting challenge to build a radio *completely from scratch*. It would be hand built, without special tools or materials, using technology that would be more-or-less consistent with early days of radio. If a book or plans for such a project existed, I had never seen it. I figured it might be time to write one.

My ground rules were simple. I would allow myself the "base" technology associated with the purchase of nuts, bolts, screws, and sheet metal. I would also allow myself to salvage certain materials from contemporary junk. However, anything that I used would have to have been available in the 1920's. No commercial electronic parts of any kind would be allowed.

Since wire is no longer available with cotton insulation (as it was in the past) I permitted myself to use enameled magnet wire, and salvage it as necessary from contemporary scrap. I also allowed myself to use certain modern incidentals, if they did not grant me an unfair advantage. So, I made free use of polyurethane instead of shellac, and modern glue instead whatever was available back then.

I limited my use of power tools to a common electric hand drill.

The electric drill made my life a little easier, though I didn't use it for any task that couldn't have been done by hand. I allowed myself the use of an electric soldering iron, and in a couple of instances, a propane torch.

Finally, it was my intent from the beginning to share my fun, and make my equipment reproducible. Therefore, I limited myself to those materials that I presumed to be universally available. As you might well imagine, these rules posed significant limitations.

If you've played with crystal radio before, you are aware that certain components are traditionally considered "purchase" or "salvage" items. Nobody attempts to manufacture them at home.

Headphones are a good example of this. As you may know, a crystal radio is useless without a sensitive pair of high impedance headphones. Yet, they are difficult to come by. Conventional wisdom says that it's impractical to fabricate them at home. At the same time, they're almost impossible to find for purchase. You can't buy them, you can't make them, and you can't salvage them. (They're not "junk" anymore -- they're antiques!) I have read the work of more than one 1990's crystal radio enthusiast who listens with his 1940's era head set.

Finding tuning condensers poses similar problems. When I was a kid, there were plenty of old tube radios at garage sales that you could purchase for a buck and then tear apart. The radios that have survived are now assuming "collector" status. In any event I don't think tuning capacitors are mass produced by anyone anymore. For the most part, modern tuners are completely electronic. Like headphones, tuning condensers tend to be the type of component that the average experimenter cannot fabricate himself.

How do you reconcile an ambitious goal with seemingly impossible limitations? How do you embrace limitations in materials or technique and fabricate components that common wisdom says cannot be built? The thinker can.

Identifying The Essential Characteristics of a Solution

I recall when I was young, a fair-sized pine tree that grew adjacent our house. As a sapling, it had been planted there without regard to the fact that it would someday become much larger. It grew to the point where it's branches beat against the siding when the wind blew, and sap issuing from the needles left stains. Clearly, the tree had to go. My father was sad about the prospect of cutting down such beautiful tree, and instead decided that the tree could be moved.

What would you need to accomplish this task? What sorts of tools or machinery? How much time and how many men? Would it surprise you to learn that he executed his plan *by himself*? I challenge you to pause for just a moment, and see if you could devise a scheme to do the same, by yourself, with no more horsepower than the strength of your own back.

My father constructed a tripod of some stout timbers. The legs of the tripod were joined, loosely at the top, probably with some heavy eye bolts. The tripod was erected next to the tree. The tripod was probably half the height of the tree. A sling was wrapped around the trunk at about one-third of it's height and a block and tackle coupled the sling to the peak of the tripod.

Next, my father, brother, and I began shoveling around the base of the tree to detach it from the earth, while at the same time leaving a ball of earth around the tree's roots. As we dug, my father would increase the upward tension on the tree by advancing the block and tackle. We had to sever some large roots, and work our way past some large stones, but the tree finally came loose. It was hoisted further so that the earth-ball was suspended a few inches above the ground. I should also mention that during this time, we had decided on the new location for the tree, and had dug a suitable hole in which to plant it.

Carefully, moving just one leg of the tripod at a time, my father actually *walked* the tree to its new location. This is not only incredible, but perfectly logical. Whenever he moved one of the legs, the bulk of the tree's weight was born by the other two legs! It took some time, and a lot of effort, but the tree was eventually moved several yards, and lowered into the new hole. It remains there to this day.

What equipment did you specify when I asked you to plan a tree move of your own? I would bet that you probably considered a crane or a large backhoe. Most reasonable people would. The problem with "standard" answers is that they tend to come from *memory*, not from the *analytical* portions of your mind. You've been taught that a crane will lift heavy things, and perhaps you've seen one in operation somewhere. Does that make it the solution to every lifting problem? By answering yes, you turn a blind eye to a thousand better solutions. The creative and innovative answer requires less recall of the past, and more analysis of the present.

What plugs into a wall socket? A television? A lamp? In truth, the wall socket doesn't care, as long as your appliance has certain essential characteristics. A wall plug must have two flat metal blades of the proper size and length. The blades must be supported by an insulating material, and the blades must be attached to a wire to carry the current to your device. The device must be capable of running on house current, and must not consume more than a certain amount. If your appliance meets these essential characteristics, exactly what it does is immaterial.

If you treat a problem like a metaphorical wall outlet, you can look upon any solution as a black box. You determine the minimum requirements that allow the solution to "plug in" to the problem. Any idea that matches the "plug" characteristics is, by definition, a solution! Once you look at things this way, the solution either becomes obvious, or your attention will be turned to a whole class of solution-candidates that you might otherwise have overlooked.

This is precisely the thought process I employed in the construction of the components in this book. In each case, I attempted to identify the essential characteristics of the generic component. Then, I combined that analysis with my materials and self-imposed limitations, and the result was fully functional equipment.

You will note that the "projects" in this book are not complete radios but components. This approach was taken on purpose. First, I wanted each part to stand on its own, for the sake of clarity. Each component is based on different principles, and each required it's own consideration. As examples of creative thought, the components best serve this purpose as separate entities.

Second, I like the idea of separate "laboratory" style parts that can easily be mixed and matched to create various radio "hookups". If you stumble upon a favorite hookup, you can always replicate your lab components and assemble them as parts of a complete and permanent radio set.

In the introduction, I alluded to the aesthetic aspects of crystal radio. If your projects are functional, they need not be pretty. On the other hand, I've argued that sets of this type have character. Why not embrace that? They are meant to be looked at, handled, and tinkered with. A good radio set that also looks good is a source of double enjoyment.

You will note that I invested considerable effort in the appearance of most of my equipment. I like clean lines and "warm" materials, like stained wood, and polished brass. I guess what I'm trying to say is that there is nothing wrong with putting a little art and a little piece of yourself into your creations.

Chapter 3
Safety First

Recently, I stopped in a drugstore to buy a greeting card for someone. Scanning the display of cutesy panda-bear and flowered birthday cards, my eyes were instantly drawn to a plain card with the word "SEX" printed in large letters. Of course I picked it up, and opened it, not really knowing what to expect. Inside, the card read: "Now that I have your attention, Happy Birthday."

Given the dry nature of the average safety lecture, it occurred to me that this tactic might have application elsewhere. Yes, I had given serious consideration to titling this chapter: "SEX". Nothing is more important than your safety, and I wanted your complete attention. Before we head to the workshop or the lab, let's remember a few important guidelines:

Always protect your eyes! Safety glasses are cheap insurance; your eyes are irreplaceable. Anytime you saw, drill, or solder, your eyes are at risk. Paint or solvents can burn your eyes. Safety glasses are a good idea anytime you are using tools, either on the bench, or in the yard.

Protect your ears! Saws and drills can make a lot of racket. Wear ear plugs anytime your ears are exposed to loud noises. Good hearing is priceless, and an important asset in crystal radio work. Don't forget to wear your plugs when you mow the lawn, and take it easy with the boom-boxes, car stereos, and stereo headphones.

Keep your focus and attention when you work! Anytime you work with tools, or for that matter, electricity, you need full use of your faculties. Don't drink, eat, or smoke anything that would impair your judgement. Don't work where you are liable to be distracted. Accidents are called "accidents" because you aren't expecting them.

Watch out for power lines! You may consider stringing up an antenna for your radios. Do not string an antenna near *any* power line. Contact is lethal. Exercise caution if you use a ladder.

In this age of litigation, it is tempting to try to anticipate every possible mistake a person might make, and issue the appropriate warning. That is a game that no author can win.

Please, use your own common sense. **You** are ultimately responsible for what you do, and responsible for your own safety. If something an author suggests does not seem safe to you, ***don't do it.*** If some technique requires more skill than you possess, ***don't do it.***

If you are one of those individuals that God blessed with two left hands, please be content to enjoy books like this one for their academic value, and forget about the tools.

Make your projects memorable for the joy that they brought, not for the time you spent in the emergency ward.

Now, let's start talking radio...

Chapter 4
Basic Theory

To experiment with crystal radio, it pays to have a basic understanding of what our equipment is supposed to do and why. A little bit of theory is in order.

A chapter like this always poses something of a dilemma for an author, because of the diverse background of the intended audience. Is this the first radio book that you've ever picked up, or are you a trained engineer? I'm going to assume, for the moment, that you are among the former. In any event, it's probably better to repeat what you may already know, than omit something that you don't. Experts, please feel free to fast-forward to the next chapter.

For the rest of us, let's begin by considering a common tuning fork. If you strike the fork, putting energy into it, it begins to ring. No matter how you strike it, it always produces the same note. Why? The tuning fork is said to be mechanically tuned or *resonant* at that frequency. Exactly what frequency the fork rings at is dependent upon the mass of the fork, the stiffness of it's tines, and other factors.

We hear the note produced by the fork because the fork disturbs the air and sends a series of pressure waves radiating in all directions. If we installed a sensing device at some fixed distance from the fork, we would detect the air pressure rising and falling, hundreds of times a second, as the waves passed our sensor. If we plotted those pressures as a function of time, we'd end up with a graph that looks something like Figure 4-1.

One rise in pressure, followed by a drop in pressure is called one *cycle*. The frequency of the fork's sound is defined as the number of those cycles that occur in one second's time. A fork that produces the musical note "A" generates 440 pressure cycles per second. Instead of

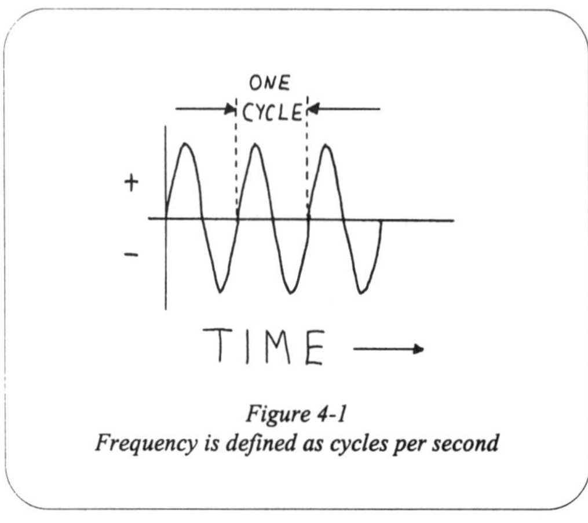

Figure 4-1
Frequency is defined as cycles per second

saying "440-cycles-per-second" which is a mouthful, engineers will instead say "440 Hertz". A "Hertz", which is abbreviated "Hz", is defined as *one cycle per second*.

A tuning fork that is struck lightly will still resonate at the same frequency of course, but the motion of the fork and the pressure waves produced will be much smaller. Conversely, if we strike the fork very hard, it's motion is violent and the waves it produces are much larger. The "height" of a wave is called its *amplitude*. Think of some ocean waves on a clear day, compared to the waves on a stormy day. Waves whipped up during a hurricane have a much larger amplitude than waves produced on a sunny day, even if they strike the beach with the same frequency. (Figure 4-2)

Here's another experiment for you: Take two 440 Hz tuning forks but strike only one. Hold the silent fork a foot or so from the ringing fork and listen carefully. The second fork will start to ring on its own! 440 Hz waves leaving the first fork and traveling through the air cross the second fork. Because the second fork is also resonant at 440 Hz, it easily absorbs energy from the waves in the air and begins to ring on it's own. (Figure 4-3)

Repeat the two-fork experiment, and strike a 440 Hz fork. This time, use a 1000 Hz fork for the silent one. You will discover that because

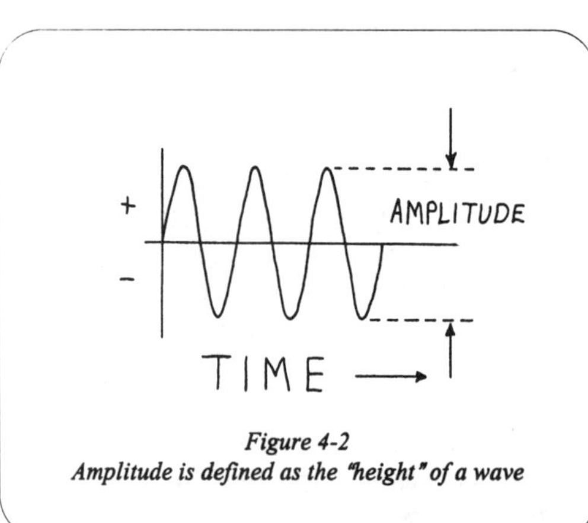

Figure 4-2
Amplitude is defined as the "height" of a wave

the second fork is "tuned" to a different frequency than the sounding fork, it's not sympathetic to the 440 Hz waves that pass it. It's unable to absorb energy from them, and remains silent. (Figure 4-4)

As it turns out, we can build an electronic equivalent of a tuning fork by using two electrical

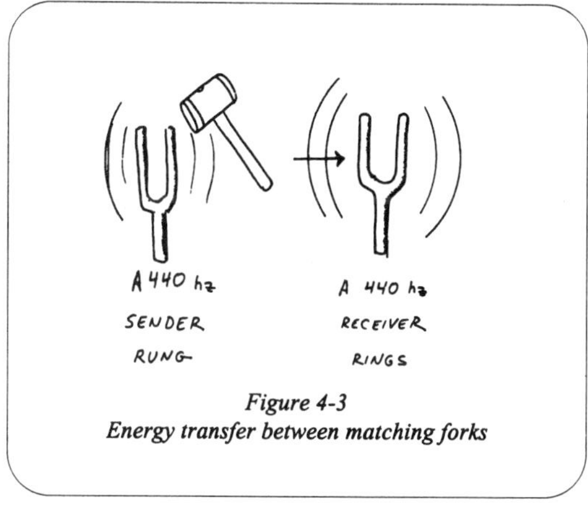

Figure 4-3
Energy transfer between matching forks

components, a *coil* and a *condenser*. Together, they form the basis of a circuit that will resonate at, or "ring" at a specific frequency. The coil, which is typically composed of loops of wire is represented symbolically by curly line. The condenser, which is typically composed of two metal plates separated by an insulator, is represented in diagrams by two short parallel lines.

The coil/condenser combination creates what is called a *tuned circuit*. (Figure 4-6) The tuned circuit exhibits most of the characteristics of the tuning forks we discussed earlier. If a tuning fork is struck with a mallet, it will ring. If the tuned circuit is "struck" with an electrical impulse, it will also "ring", (though you can't hear it!)

A ringing fork radiates sound waves. The tuned circuit also radiates waves, though the waves are electromagnetic. Electromagnetic waves are related to light, and travel at the same speed as light. Unlike sound waves, they don't require a medium to travel through. In fact, the vacuum of deep

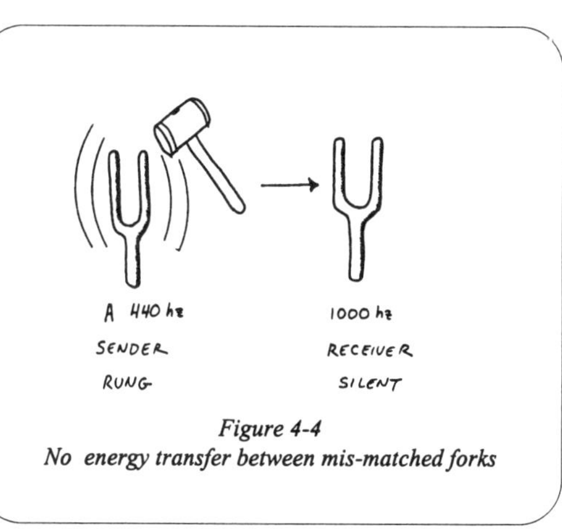

Figure 4-4
No energy transfer between mis-matched forks

Page 13

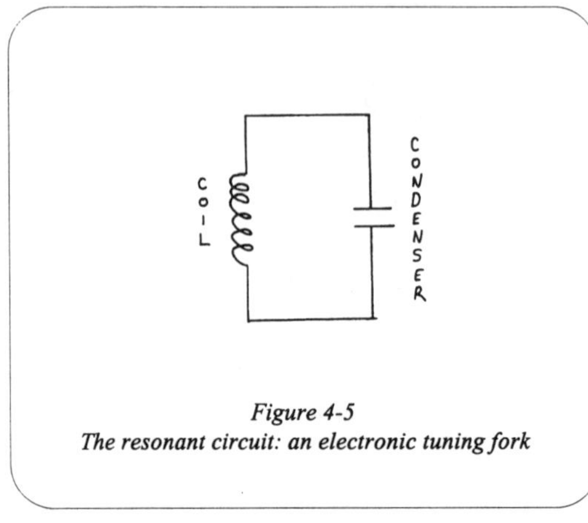

Figure 4-5
The resonant circuit: an electronic tuning fork

space suits them just fine. Radio waves tend to have much higher frequencies than sound waves. As mentioned earlier, the "A" note on a common tuning fork has a frequency of 440 Hz. Human hearing goes up to 20,000 Hz.

On the other hand, your favorite A.M. radio station has a frequency on the order of a *million* Hz. The frequencies used in cell phones, operate at around 900 megahertz, or nine-hundred-*million* cycles per second! Needless to say, you can't hear, see, smell, or touch radio waves.

Recall that a "cycle" of a sound wave involves repeating, rising and falling pressures. The cycle begins at normal pressure, rises to a positive level, falls to normal, then to a negative value, and finally back to normal. Electromagnetic waves begin with a nominal electrical voltage, which rises above and falls below a nominal value, in repetition.

If a silent fork is exposed to the waves generated by another fork, and both forks are set up to resonate at the same frequency, the silent fork will capture some of the transmitted energy, and will begin to ring itself. Tuned circuits do the same thing. They can capture energy from passing electromagnetic waves, provided that the circuit is tuned to resonate at the same frequency as the waves that strike it.

The earliest wireless telegraph systems consisted of little more than the

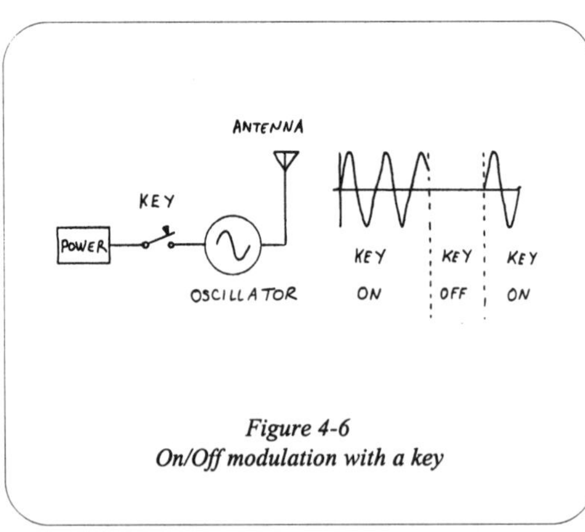

Figure 4-6
On/Off modulation with a key

circuits we've just discussed. In general, a telegraph key was wired so that, when it was pressed, it would stimulate a tuned circuit to ring (*oscillate*). Often, high voltage and electrical sparks were used to "strike" the circuit and induce it to ring. When it rang, it emitted radio waves.

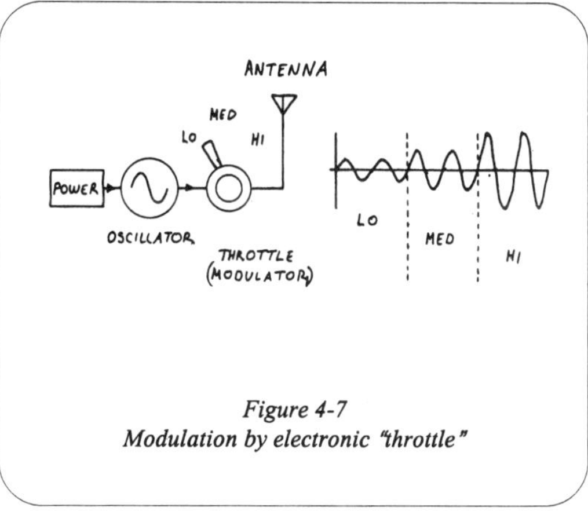

Figure 4-7
Modulation by electronic "throttle"

When the waves reached the receiver, another tuned circuit captured some of the transmitted energy. Because the captured electrical currents in the tuner were so small, they could not directly activate the telegraph's sounder. A device called a coherer was used to sense the feeble currents in the tuner, and activate the telegraph sounder.

In the wireless telegraph, the transmitted wave is called a *carrier*, because it literally "carries" our message. The intelligence in the message is implied by the presence or absence of the carrier. In other words, if our receiver captures a carrier, this means that the telegrapher at the transmitter just pressed his key. If the carrier vanishes, we assume that the sender's key was released. Using morse code, we can use on/off patterns to represent letters and words. (Figure 4-6)

Are there other ways to encode information on a carrier? Absolutely. Suppose that the key was eliminated, and the transmitter was modified to include an electric "throttle", so that you could vary the amplitude of your carrier by adjusting a

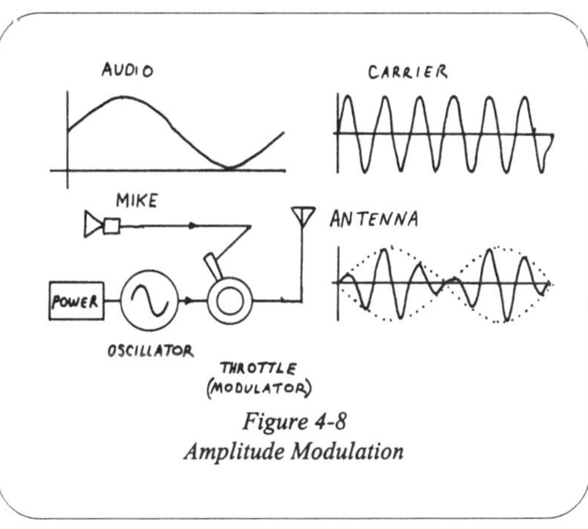

Figure 4-8
Amplitude Modulation

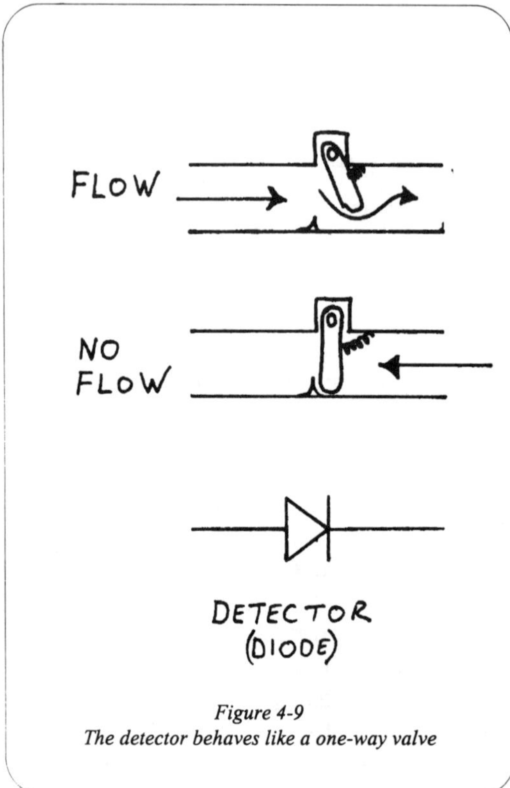

Figure 4-9
The detector behaves like a one-way valve

lever. By throwing the lever one way or another, you could make your carrier on, off, or anywhere in between. If you assigned meaning to the different levels, the person receiving your transmission could monitor the amplitude of the carrier that his tuner captured, and decode your message. Throttling a carrier in this way is called *amplitude modulation*, because you are modulating, or changing, the amplitude of your carrier to encode your message. (Figure 4-7)

The ultimate refinement is to rig the modulator (throttle) to a microphone, to allow the tones in your voice to modulate the carrier up and down. A loud tone would allow the carrier to be transmitted with little attenuation. A soft tone would cause the carrier to be throttled way back. In essence, the modulator "shapes" the carrier to reflect the general shape of the microphone signal that modulates it. Figure 4-8 shows a block diagram of the complete process, and there you have it: an A.M. transmission. All we have to do now is receive and decode the carrier in order to reconstruct the original message.

We begin with the same tuned circuit as before. The coil and/or the condenser are chosen so that the circuit will resonate at the desired carrier frequency. This allows the tuner to capture energy from the transmitted signal.

Extraction of the sound signal is done with a device called a *detector*. The detector is like a check valve or a one-way turnstile. It

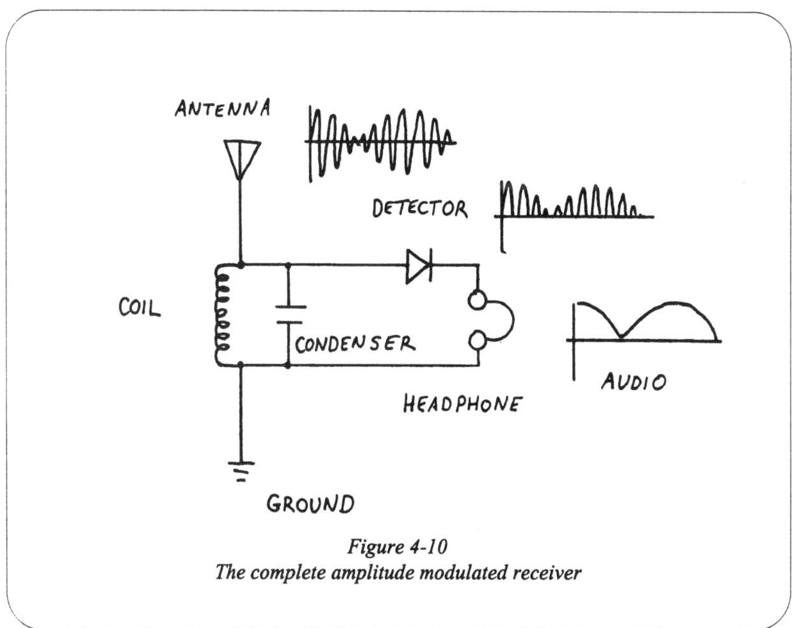

Figure 4-10
The complete amplitude modulated receiver

allows current to flow in one direction, but not in the other. The detector symbol is drawn like an arrowhead stuck into a small plate. (Figure 4-9)

If we feed the captured signal through a detector, only the top half of the signal is allowed to pass. (The bottom half is thrown away.) The result is a wave that looks a lot like the original sound wave, except that it still contains the high-frequency components of the carrier.

As it turns out, this is not a problem. The high-frequency component can easily be filtered out. This leaves an electrical signal that represents the original sound. If that signal is now fed to an appropriate speaker, the electrical signal is converted back into sound.

Since speakers are mechanical devices, and mechanical devices are normally not responsive to high-frequencies, it is often possible to feed an unfiltered, detected signal directly to the speaker. The speaker simply ignores what's left of the carrier. Figure 4-10 shows a classic circuit for a crystal radio. It embodies everything that we just discussed. Note the addition of an antenna and a ground, which aids the tuned circuit in gathering energy.

Chapter 5
Essential Characteristics Of A Good Headphone

There is probably no more crucial a part of a radio set than the speaker or headphone. You can manipulate electronic signals all you want, but until you can convert those signals back into sound, your radio may as well be a cinder block. Ironically, it's one of the hardest to find components, which is good reason to try to design your own.

How might we go about building a set of headphones? First, we have to identify the essential characteristics of a "good" set of headphones. Much of what makes headphones either "good" or "bad" is based on the idiosyncracies of the crystal receiver itself.

Remember that the crystal set is passive; it does not amplify. In terms of signal strength, what you receive is what you get. It stands to reason, then, that all crystal radio components should be as efficient as possible. A good speaker or headphone is able to produce ample volume from tiny signals.

Yet, efficiency alone does not define a good headphone. Let me pose an example. Suppose you seat yourself in front of an (unplugged) table fan and a toy pinwheel. The table fan is a machine built to fairly precise tolerances. It probably has computer-designed blades and excellent bearings.

The pinwheel, on the other hand, is made of very cheap materials. The curvature of its blades has less to do with airflow than with aesthetics. It's bearings are probably no better than a pin driven through the folded corners of the blade. In terms of mechanical efficiency, the pinwheel is crude at best.

Now take a deep breath and blow, both at the pinwheel and at the fan blade. The pinwheel spins vigorously, but you can get light-headed trying to make the fan blade move. Why?

The difference lies in *loading*. The fan blade is large, and relatively heavy. It takes a good bit of wind to move it, and thus represents a significant load (to you!) The pinwheel, on the other hand, is small, and light. It represents such a small load that it spins easily, even though it wastes energy in it's poor blades and in it's poor bearings. The moral of the story is this: the efficiency or sensitivity of your headphones is important, but may not be as important as how much it loads the receiver.

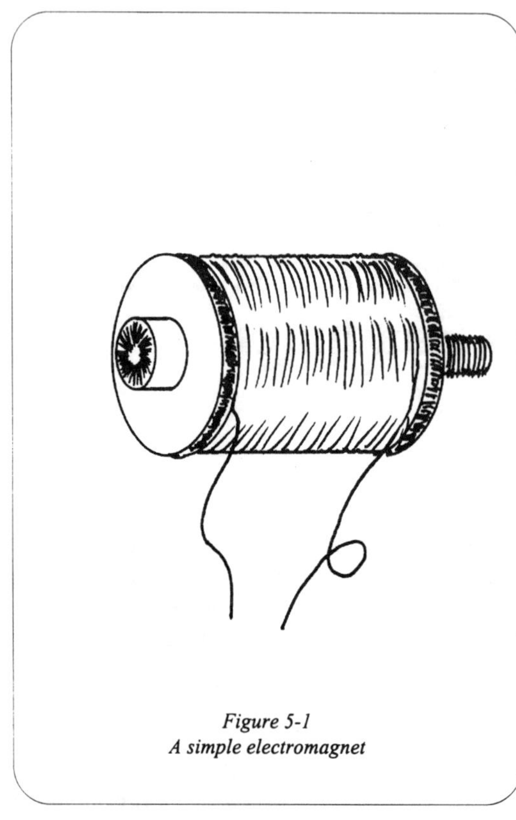

Figure 5-1
A simple electromagnet

A headphone's tendency to load a receiver depends on it's *impedance*. Impedance is measured in ohms and represents the extent to which the headphone resists the flow of current through it. High impedance headphones allow only small currents to pass through them, which results in very little loading. Low impedance headphones require larger currents, and they can cause significant loading on the receiver.

Most crystal radio projects I've seen call for headphones with 2000 ohms or greater impedance. The average stereo speaker, by comparison, runs only 8 ohms. I've never seen headphones for portable cassette players run much above 30 or 40 ohms. Compared to the "good" headphones, modern speakers and headphones represent a load that is several *hundred* times greater. They are essentially unuseable for crystal work, despite their excellent efficiency.

Any one who has ever tinkered with electricity has probably taken an old steel bolt, and wrapped wire around it to create an electromagnet. If you apply a voltage to the coil with battery, the bolt will become magnetic. When you remove the power, the magnetism vanishes. You may also have noted that the greater the voltage applied to the coil, the stronger the magnet you produce. (Figure 5-1)

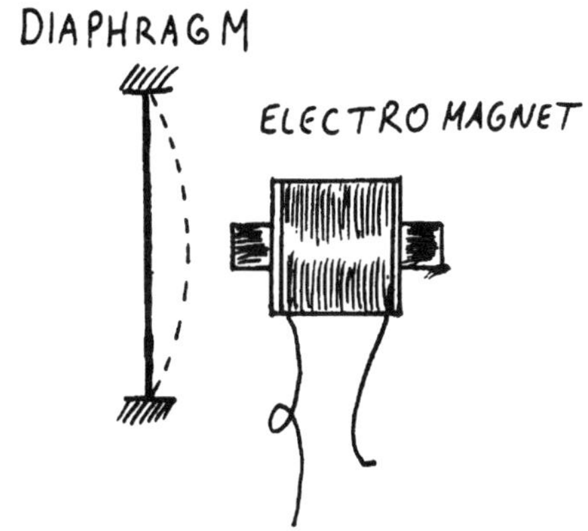

If we place just such an electromagnet in close proximity to a thin sheet of steel or iron, and feed the coil an audio signal, the coil's field will become weaker and stronger as the audio signal fluctuates. This will cause the sheet metal, or *diaphragm*, to be attracted to a lesser or greater degree, and it's vibrations will produce sound. This is the basis of a simple electromagnetic speaker or headphone.

This primitive design works, but if you think about a single cycle of audio, you'll see that the motion of the diaphragm doesn't always track the signal properly. The cycle begins at zero voltage. The coil in our simple speaker produces no field, and the diaphragm is at rest. As the cycle progresses, the voltage increases, and so does the magnetic field, so the diaphragm is drawn inward. Next, the voltage returns to zero, and the diaphragm relaxes. (Figure 5-2)

The problem lies with the last portion of the cycle. As the voltage goes negative, ideally, the diaphragm should move *outward*. Of

Figure 5-2
Movement of a simple diaphragm

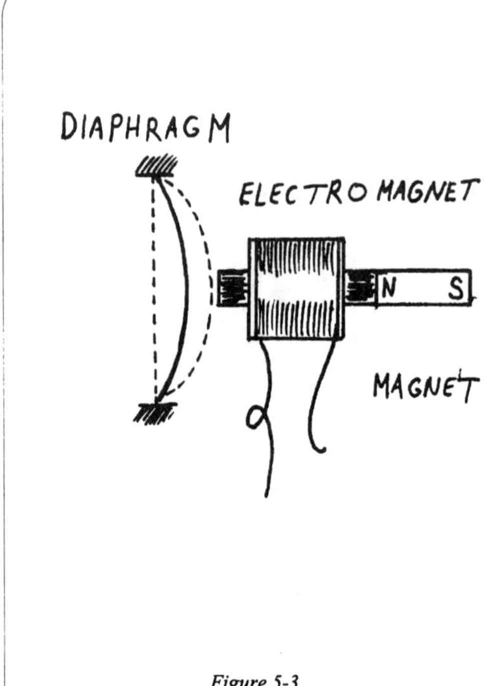

Figure 5-3
Movement of a diaphragm biased with a permanent magnet

course, it does not. It moves inward again. As I said, the diaphragm fails to move in conjunction with the audio signal driving it. Because the diaphragm will only move in one direction from its resting state, the sound produced is not nearly as loud as it would be if the diaphragm was able to move in both directions.

We can materially improve the response of the diaphragm if we apply a permanent magnet to the coil's core. Think about the same audio signal again.

At zero voltage, the coil is dead, but because of the permanent magnet, the diaphragm is drawn inward just a bit. This is now the resting position. As the audio signal goes positive, the coil is energized and produces it's own field, which adds to the field already in the bolt. The diaphragm is drawn inward even farther. When the audio cycle returns to zero, the coil's field disappears, and the diaphragm relaxes to the "rest" position where it is held by the permanent magnet.

When the audio signal goes negative, the coil becomes energized again. This time, however, the magnetic field produced by the coil is in opposition to the field produced by permanent magnet. In combination, the two magnetic fields cancel each other. The diaphragm moves from the resting position, *away* from the bolt.

As you can see, by combining the fixed field of a permanent magnet with the variable field of a coil, we can get a metallic diaphragm to vibrate over a full range of motion. The sound it produces more

accurately represents the audio signal that is fed to the coil. Because the diaphragm moves twice as far, the sound produced is much louder. (Figure 5-3)

There are other points of good design to ponder. For instance, magnetic fields weaken dramatically over distance. Therefore, if we want to maximize the influence of the electromagnet on the diaphragm (to maximize sensitivity), we want the core of our coil to be located as close as possible to the diaphragm without actually touching it.

We said that the diaphragm is fashioned from sheet of a magnetic material like iron or steel. Obviously, it's easier to distort the shape of a thin diaphragm than a heavy, thick one. We want the diaphragm to be able to respond to subtle fluctuation in the coil's magnetic field. The diaphragm needs to be pliable and light.

In our conceptual speaker design, we discussed using an old bolt as the core of our electromagnet. Obviously, any core must be a magnetic material like iron or steel. The catch is that there are many types of steel, and variations in iron as well. Some of these materials, when wound with coils, produce excellent magnets. Others, using the same coil, produce very poor magnetic fields. Your headphone is only as sensitive as the quality of your electromagnet. We need to be very picky about the materials we use in the core.

One of the unfortunate side effects of passing a changing current through any coil wrapped on a metal core, is that the core tends to warm up. The changing currents in the coil actually induce currents in the core itself, which the core converts to heat. This is wasteful. It means that some of the energy you put into the coil was squandered as heat, instead of doing whatever you wanted it to do. Large cores seem to waste more energy than tiny cores.

One of the strategies used to combat this type of loss is to laminate the core. Instead of winding a coil on a square rod of steel, for example, the coil is wound on a stack of thin steel sheets. In the end, the laminated core is the same size and shape as if it were solid. But, because it is composed of many smaller cores, the total core loss is much less. For the sake of efficiency, a laminated electromagnet core is probably a good idea.

Before I show you my first design, let's review the essential characteristics of a good set of crystal radio headphones.

We said that our design must be as efficient as possible in converting audio signals to sound. The voice coil should have as high an impedance as possible in order to minimize loading on the receiver. The voice coil should also measure 2000 ohms or greater. It should be wound on a laminated core made from a material that is highly susceptible to magnetism. The diaphragm has to be metal, and magnetic, yet thin and light. Don't forget the permanent magnet in order to bias the core and enable the diaphragm to move in both directions.

Chapter 6
The "Gallows" Headphone

Figure 6-1 shows the first headphone design that I'd like to share with you. It was not modeled after any particular piece, but evolved into it's present shape based on the tools and materials I had on hand. After building this device, I was amused to discover drawings of some ancient telephone receivers that bear a striking similarity to my device.

The old units are referred to as "gallows" style receivers because of their resemblance to the frame of the old hangman's gallows. This resemblance is less apparent with my speaker, but the name is as good as any.

To operate this device, you need a doctor's stethoscope. These can be purchased new for as little as six dollars at your local drug store. The chest piece is removed from the stethoscope, and the open rubber tube is slid onto the brass pipe protruding from the front of the my instrument. Two wing nuts provide terminals with which to make electrical connections. Let's run through this design, while I explain each part, and why I made the design decisions that I did.

Let's begin with the diaphragm. I was unable to locate steel or iron sheets thin enough to serve as a decent diaphragm. However, an assortment of brass shim stock was available at my local hobby store in thicknesses down to 0.001". I wanted to use as thin a diaphragm as possible, but discovered that the real thin stock is rather fragile. I settled on the 0.003" material. Of course, brass is not magnetic, so you may wonder how such a material could possibly work — by cheating, of course! I cut a dime-sized piece of steel, and soldered it to the center of the diaphragm. This would be the region acted upon by the voice coil.

I used a shoe polish can as a sound chamber. The top of the

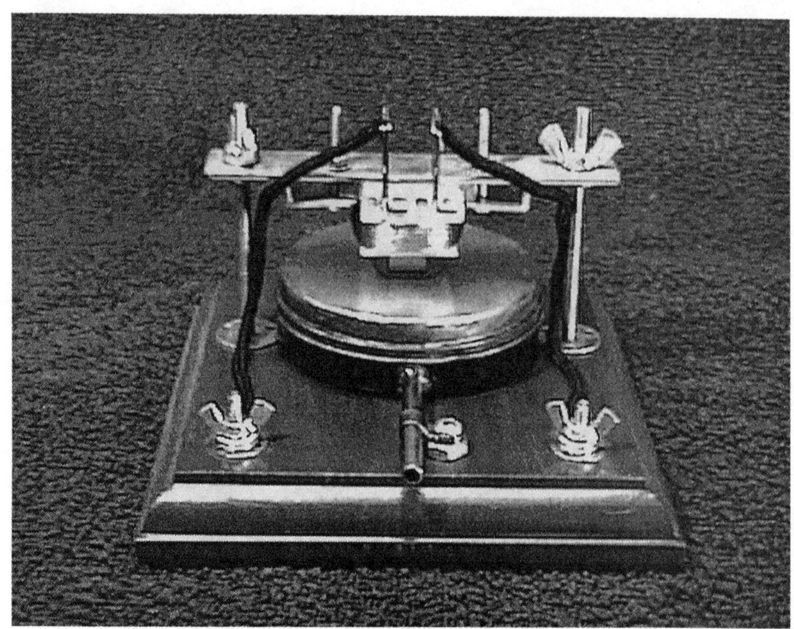

Figure 6-1
The "gallows" headphone

can's lid was cut out, leaving about a 1/32" rim. The rim was sanded to remove the paint, and the brass diaphragm was soldered into place. The bottom half of the can features a metal twist-tab which is used to open it. I drilled out the rivet and removed the tab. I enlarged the hole with another drill, and then soldered a 3/8" brass tube into the hole. (Figure 6-2)

I drove sheet metal screws through the bottom of the can to attach it to a wooden base. The can's lid, with it's diaphragm, was replaced. I attached a stethoscope to the brass tube and listened. Any movement of the diaphragm produced plenty of noise in the stethoscope.

Next, I turned my attention to the core of the voice coil. I could have cut strips of sheet metal and stacked them to form a laminated core, but this seemed like a lot work. Instead, I salvaged the laminations I needed from a discarded transformer. (Figure 6-3)

The transformer was removed from one of those black plastic wall adapters, commonly used to power small radios or answering machines.

First off, I knew that whatever the laminations were made of, it was selected by the transformer's manufacturer for it's good magnetic characteristics. Second, I found it easier to modify the salvaged laminations than cut my own. The laminations from the transformer were shaped like the letter "E". Using tin snips, I cut off the outside legs to produce the letter "T".

Next, came the voice coil itself. I could have fabricated a cardboard spool to fit the "T" shaped core, and then wind the coil myself. Instead, I recycled one of the plastic spools that had been part of the scrapped transformer.

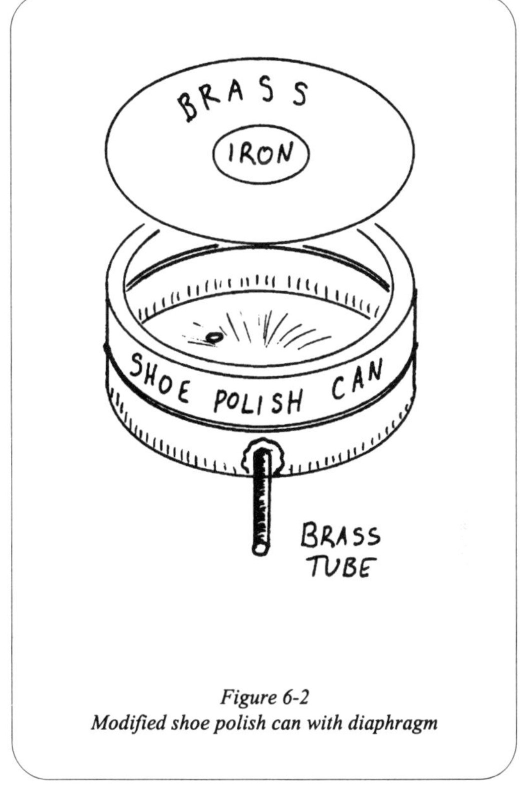

Figure 6-2
Modified shoe polish can with diaphragm

Wall transformers typically have two windings. One, made of very thin wire, is connected to the prongs that plug into the wall. This is called the *primary coil*. The *secondary coil* is made of thicker wire, and it delivers power to your appliance. I kept the spool that contained the fine primary wire. This gave me a spool that not only fit my core perfectly, but was already wound to boot.

Using an ohm meter, I measured the impedance of the coil and found it to be far short of my 2000 ohm target. I scrapped another transformer to salvage it's primary wire, and spliced it to the wire already on the spool. I added turns to my coil until no more wire would fit onto the spool. At that point, the coil read 1000 ohms. I figured I'd go with that for now. The spool was slipped on to the core, and painted with a few coats of polyurethane. This makes the coil stronger and more

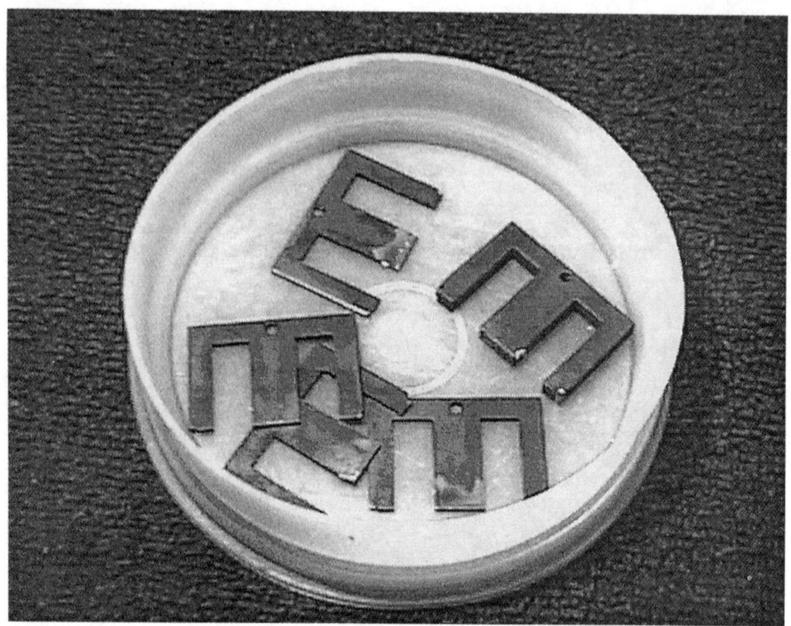

Figure 6-3
Transformer laminations

rigid, and also glues the laminations together.

Two holes were drilled in the base, one on either side of the shoe polish can. 1/8" threaded rods were anchored in these holes with nuts and washers, and cut to project vertically about 2 1/2". The ends of the rods were bridged with a strip of 1/16" brass stock, measuring 4 1/4" x 3/4". The rods and the brass strip constitute the frame of the "gallows".

The brass strip is suspended above the shoe polish can, and crosses the approximate center of the diaphragm. It's secured to the rods on each side with a hex nut and a wing nut. By adjusting the nuts, it's possible to raise or lower the brass strip to any distance from the diaphragm.

Two "L" shaped brackets were fashioned from some scrap aluminum, and used with 4-40 screws to clamp the "T" shaped voice coil core onto the brass strip. In effect, the gallows frame suspends the coil just above the diaphragm. (Figure 6-4)

Adjustments were made so that the core was aligned with the steel disk soldered to the diaphragm. Also, the hex and wing nuts were

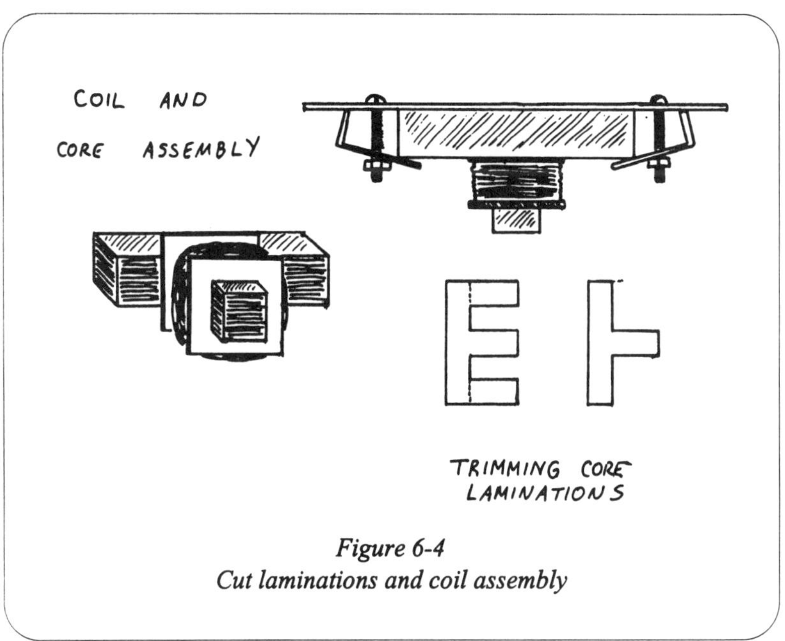

Figure 6-4
Cut laminations and coil assembly

adjusted so that the smallest possible gap existed between the face of the core, and the diaphragm.

I set the gap this way: I backed off the nuts so that the brass/coil assembly was loose. Next, I inserted a piece of heavy paper between the diaphragm and the coil's core. I carefully tightened the hex and wing nuts to lock the coil into place. When the paper was removed, only a tiny gap remained.

Figure 6-5 shows the relationship between the shoe polish can and the coil assembly. The spacing between the coil and the diaphragm is shown *before* proper adjustment.

Wait a minute! Could it be that something is missing? Didn't we say earlier that a good headphone requires a permanent magnet for clarity and efficiency? Where is the magnet in this design? Was it overlooked? No, I just applied the principle a little differently than you might expect.

In the theoretical headphone, we applied the magnet to the voice coil's core. That was arbitrary. If you think about it, the principle would work just the same if your magnet was attached to the diaphragm. Unfortunately, even a tiny magnet adds additional weight to the diaphragm and may hinder it's rapid movement.

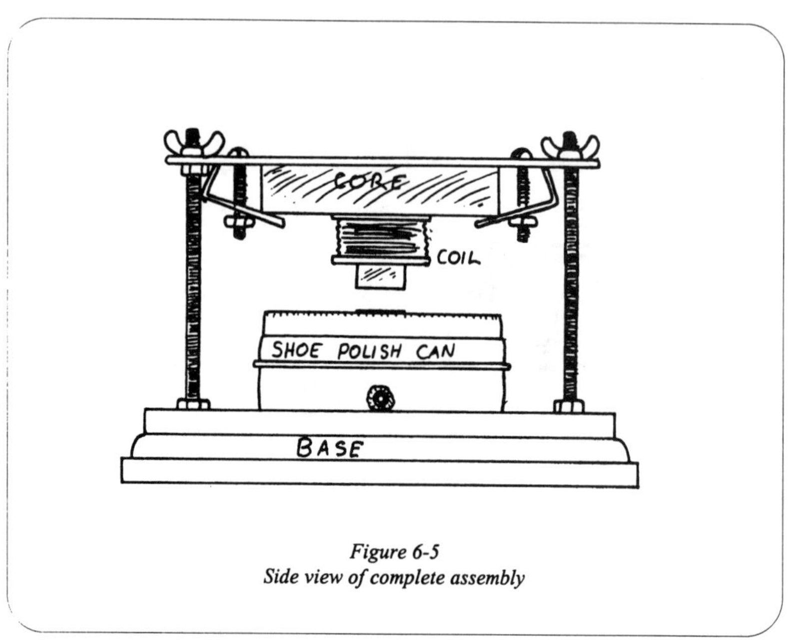

Figure 6-5
Side view of complete assembly

My solution was to place my magnet *inside* the shoe polish can, just beneath the steel disk soldered to the diaphragm. The field of my magnet radiates upward and induces a magnetic field in the diaphragm. As long as the permanent magnet rests below the diaphragm, the steel disk will also be a magnet. It's a neat approach, because the diaphragm acts like a magnet is attached to it, without any additional weight.

If you look closely at the photo of this instrument you will notice two brass terminals extending from the top of the coil. These are actually the prongs that once enabled the wall transformer to plug into the wall. To make connections to my headphone convenient, I soldered some stranded wire to the prongs, and then terminated those wires with binding posts made of 6-32 bolts and wing nuts.

To finish up, allow me to share a couple of tips with regard to aesthetics: I wanted a vintage look for the hookup wire on the headphone, but cotton covered wire is not readily available. Solution? I made my own.

I started with a package of shoe laces, selected for their round cross section (These are typically used on hiking or work boots.) I clipped the tips off of the laces, and using a pair of needle-nose pliers, pulled the core string out of the center of the laces. This leaves a hollow woven tube. It's a simple matter to thread some wire through short lengths of lace.

I also wanted the wooden base for this project to have a little class. At a local arts/crafts shop, I found a wonderful selection of milled, wooden plaques. They come in a variety of shapes, sizes, and thicknesses, all with nicely routed edges. While they are raw wood and require finishing, generally no sanding or other surface treatment is necessary.

I generally leave the wood raw while I build up the instrument. After all the parts have been located, all holes are drilled, and the instrument is complete, I then disassemble it to finish the wood. I like to use a commercial wood stain, followed by a coat or two of polyurethane. After everything is dry, I reassemble the project. A nice finishing touch is to cement a felt pad to the bottom of the base. This covers up all screw heads and gives the device a nice "feel" when it's set down onto a table top. This idea came from a genuine antique that I own.

End Results and Going Farther

This particular design is very sensitive and worth your trouble to build one. If you attach one terminal to an earth ground, and touch the other terminal with your fingertip, you can hear a pronounced buzzing noise. This is the sound of feeble currents, which radiate from the electric wires in your walls, and induce currents in your body. I have used this headphone with a variety of actual crystal sets with good success. Just the same, I have some ideas for improvements that you could make.

My instrument seems to favor low frequency audio tones, so broadcasts sometimes sound muffled. I gave some thought as to why this might be the case, and came to a couple of conclusions.

Think about a drum. If the drum skin is loose, the drum will produce lower tones than if the skin is tight. The sound chamber in this design is physically similar to a drum, with the diaphragm acting like the drum skin. A thin metal diaphragm is very sensitive, but I have no way to induce any tension in it. Thus, it stays comparatively loose, and will favor lower tones. I think it would be worthwhile to develop a mechanism for tightening the diaphragm, to select a slightly thicker diaphragm, or even a different material (plastic, paper, or mica?)

The other reason this headphone favors the low end may have to do with the voice coil itself. One of the electrical characteristics of any coil is it's *inductance*. The inductance of a coil determines it's tendency to resist changes in the currents flowing through it. The greater the inductance of the coil, the more it resists rapidly changing currents. Also, as inductance increases, the frequencies that the coil allows to pass become lower and lower. My headphone's coil may simply have too large an inductance.

Considering that power transformers are sized and wound to operate in the 60 Hz range, and considering where my core and coil came from, it makes sense that a 3000 Hz audio signal might have difficulty getting though it. I think that the fix is to reduce the size of the core.

Serviceable laminations could be cut from a steel can. Cardboard spools are easy to make, and could be custom fit to whatever core you came up with. A finer gauge of wire will allow you to get more turns onto the core, which will not only boost the impedance, but probably make the device more sensitive as well.

Chapter 7
The "Tin Can" Headphone

In the early days of radio, every radio enthusiast owned a set of "cans". I'm speaking, of course, of their headphones. I'm not sure of the specific origin of this term, but it's not difficult to see the resemblance between an old-style headset, and a pair of tuna cans lashed to the sides of an operator's head. In this chapter, I'd like to show you how to take an old buzz-word and turn it into a high-quality listening instrument.

After having built the gallows headphone, I had the good fortune to come upon an authentic, original magnetic earpiece from an old high impedance head set. I was interested to see how an original device was laid out, and what lessons it might teach me.

Unscrewing the lid, I removed the diaphragm. This was a very thin disk of what appeared to be steel. What surprised me was it's relative thickness. I pulled out a micrometer and measured it. The disk was about 0.007" thick. I began searching my scrap boxes for some sheet metal in this thickness.

In our garage we keep a box to store empty vegetable and soup cans in. Once a week, the cans are recycled. I fished some lids out of the recycle box and was amused to find several of them that also measured 0.007"! Would these make suitable diaphragms? How would I mount one of these lids? Pondering this, I absentmindedly held an empty soup can to my ear and the solution came to me: Leave the diaphragm *on the can*!

I searched in the box some more, and finally settled on mushroom can. It was selected because it met several arbitrary criteria. First, it was the same diameter as a soup can, but only two-thirds as long. I felt that this would make for a more compact device.

Figure 7-1
The "tin can" headphone

Second, in studying and comparing one can to another, I found that the number and types of creases in the lid affected it's rigidity, and thus it's sensitivity. The mushroom can's lid featured three concentric rings of varying width, depth and direction. For whatever reasons, this configuration left the lid extremely responsive to mechanical movement. This can would form the body and the diaphragm of my tin can headphone. (Figure 7-1)

Referring back to the original earpiece, I studied the coil inside. As expected, the core was laminated but comparatively small in cross section. It probably didn't measure more than 1/4" across. The core held a single coil, wound on a fiber bobbin. The wire used was inconceivably thin, as thin as my hair in fact. How could I produce a laminated core on this scale? How could I produce a coil this small, and where would I find wire like this?

Once again I searched my stock of odds and ends for inspiration. I came upon a large spool of iron wire, that I had purchased to secure some grape vines to a trellis. Iron is a good magnetic material, but this wire was too thin to use as core--or was it? It occurred to me that if I bundled several strands together, I could build up a core of any desired diameter. I also figured that a core composed of separate strands should have the same advantages as a laminated core composed of sheets.

My technique for producing the core involved gathering seven strands of iron wire, and then bundling them by winding them tightly with more wire. The outside winding was solid, and looked like a compressed spring. The winding didn't run the full length of the core, just about an inch or so. The rest of the core was left unwound, so that the strands in the center could be bent or fanned out as necessary. I fabricated three core pieces like this. (Figure 7-2)

Next, I took a junk magnet, and using a compass, identified it's poles. To the top pole of the magnet, I attached a single core piece. (Figure 7-3)

Two cores were attached to the magnet's bottom pole, and bent upward to point in the same direction as the top pole's core. This is a lot more difficult to explain than to do. Figure 7-4 should

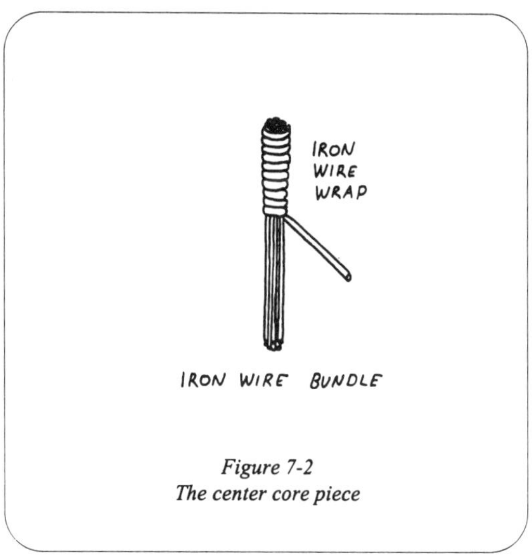

Figure 7-2
The center core piece

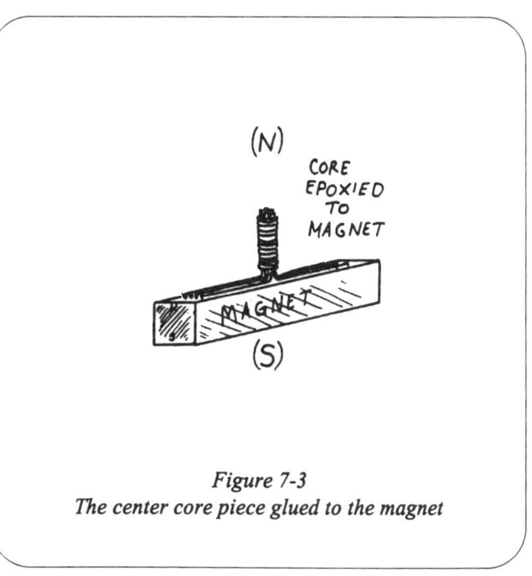

Figure 7-3
The center core piece glued to the magnet

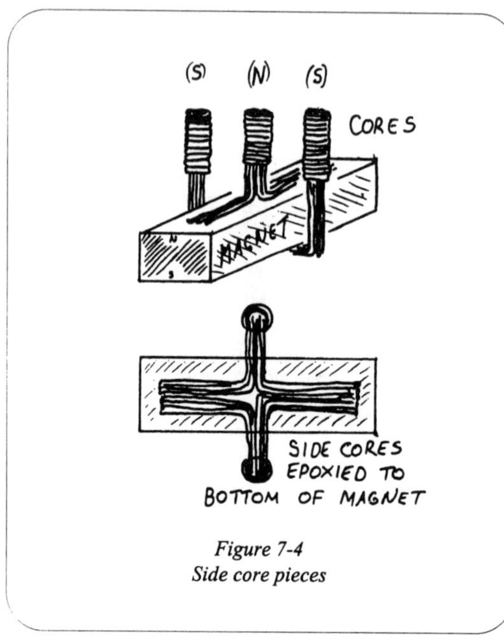

Figure 7-4
Side core pieces

clear up any confusion.

The cores were physically fastened to the magnet by fanning out their center strands, bending and trimming them as necessary, and then gluing them to the magnet with epoxy. You might think of the finished assembly as a "W"-shaped horseshoe magnet where the center leg is one pole, and the outer two legs comprise the other pole.

Using glue, a razor blade, and cardboard from a cereal box, I fashioned a small cardboard bobbin sized to slide over the center leg of the core assembly. Obviously, the spool was small enough to nestle between the outer legs of the "W". On one flange of the bobbin, near the edge, I pierced it with a pin and installed two small loops of solid copper wire (24 gauge or so). These would provide solder points at which to terminate the ends of the coil. Now, the finished bobbin needed to be wound with wire. (Figure 7-5)

The next concern was finding some wire for my coil. I happened to have

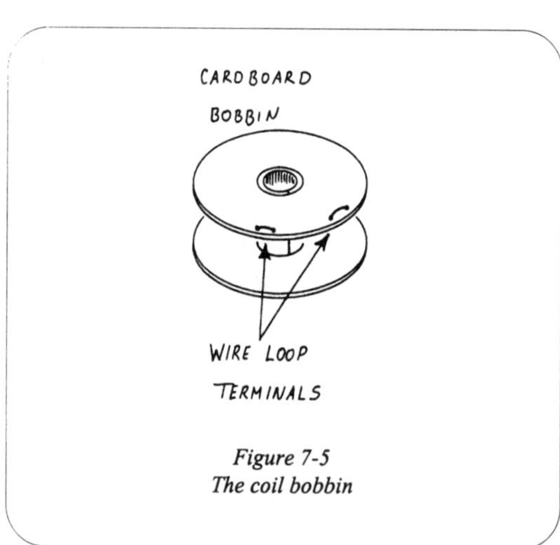

Figure 7-5
The coil bobbin

some old guitar pickups from my days as an aspiring musician. I knew them to be worthless junk, and a good source of *very* fine gauge wire . Just the same, I would recommend that you *never, never* take apart an "old" guitar pickup until you've had an expert look at it. Vintage pickups are in extremely high demand by musicians and guitar collectors alike. It would be very sad if you accidently destroyed a guitar or pickup that had historical or collector value.

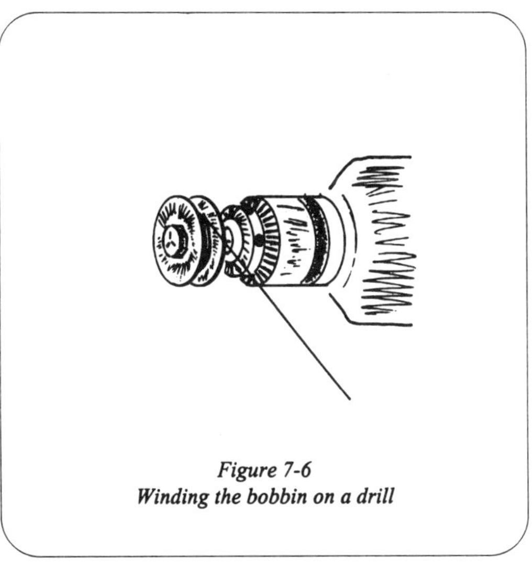

Figure 7-6
Winding the bobbin on a drill

The better approach is to visit your local guitar shop and speak to the person that does their instrument repairs. Lot's of low-end guitars are sold with very cheap (and lousy) pickups in them. Customers sometimes buy better pickups, and arrange to have them installed. What happens to the cheap pickups that are removed? They collect in a box under the counter. Sometimes a couple of bucks will free one up. Explaining your radio project may generate enough curiosity to get you an old pickup for free.

As I said, the wire inside guitar pickup coils is so fine

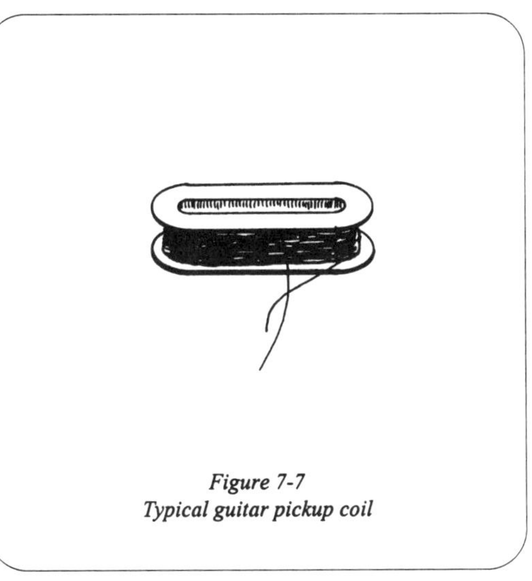

Figure 7-7
Typical guitar pickup coil

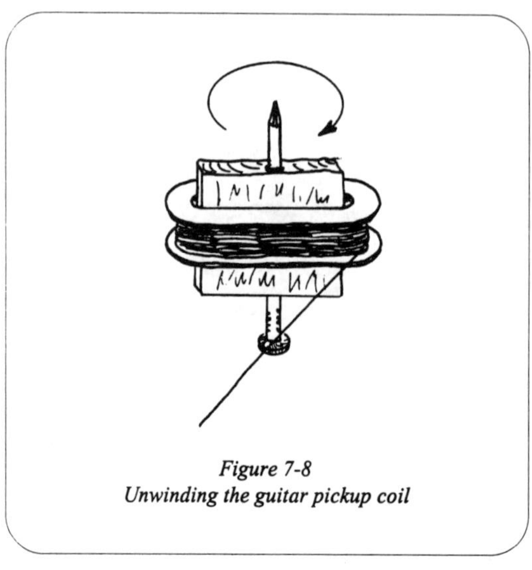

Figure 7-8
Unwinding the guitar pickup coil

that it's a little hard to work with. It breaks very easily, so anything that you can do to minimize stress on it as you transfer the wire to your voice coil bobbin is a plus. To wind my bobbin, I slipped a bolt through it and tightened it with a nut. I then chucked the end of the bolt in a variable-speed electric hand drill. My drill allows very precise control of speed, which is helpful.

(Figure 7-6)

The coil in the pickup is usually wound on a plastic spool, but the spool is oblong. (Figure 7-7) The opening in it's center is narrow and rectangular. It becomes difficult to draw wire off of the pickup's spool, because there is no good way to provide an axle for it as it unwinds. I fabricated a little piece of wood that fit into the center of the pickup's spool. Then, I drilled a tiny hole in it's exact center. I clamped a nail in my vise and slipped the spool onto the nail. It spun freely and I was able to pull wire off with little difficulty. (Figure 7-8)

Despite all of my care, I broke the wire three times! After each break, I would splice the wire with a

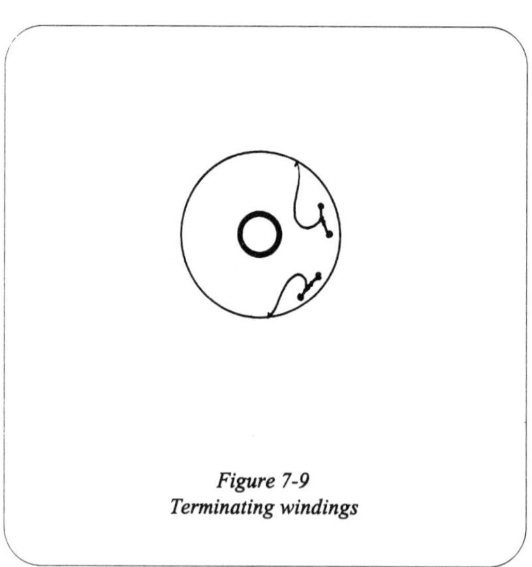

Figure 7-9
Terminating windings

tiny solder joint, and continue winding. Almost all of the wire on the pickup transferred to my bobbin. The delicate ends of the coil were soldered to the terminals on the bobbin. The finished coil was tested with an ohm meter and measured 4,400 ohms. (Figure 7-9)

I soaked the entire coil in polyurethane overnight, and then

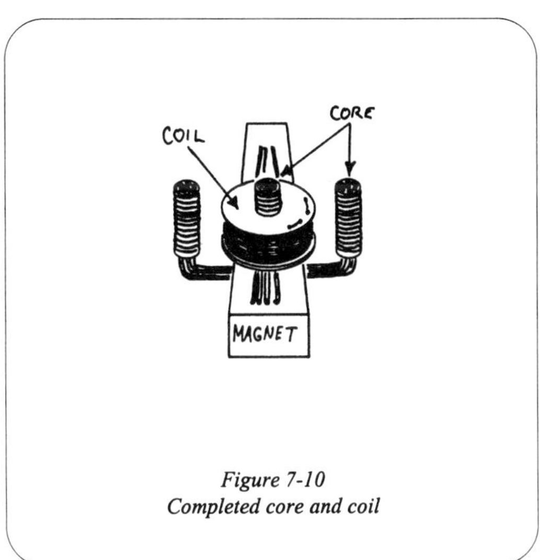

Figure 7-10
Completed core and coil

hung it up to dry. The polyurethane adds strength, both to the coil and the bobbin, and makes them more tolerant of humidity. The finished coil

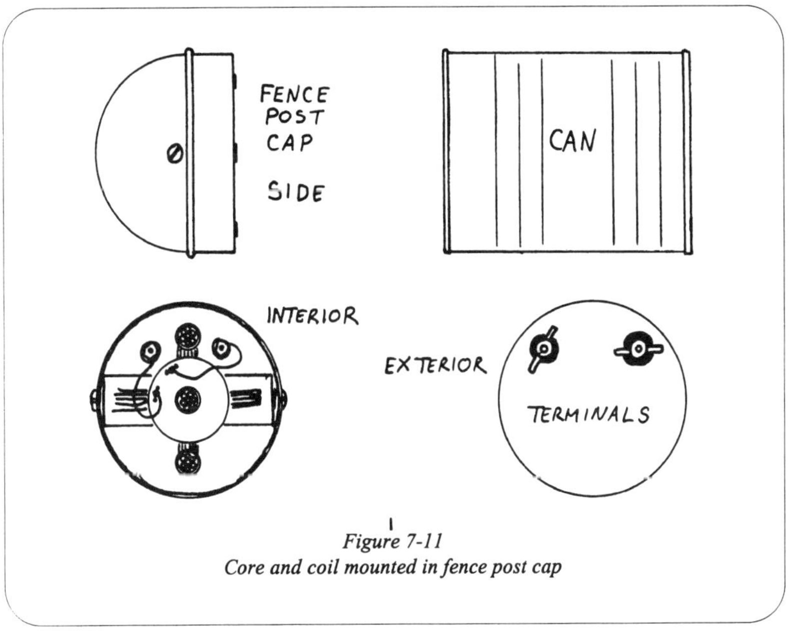

Figure 7-11
Core and coil mounted in fence post cap

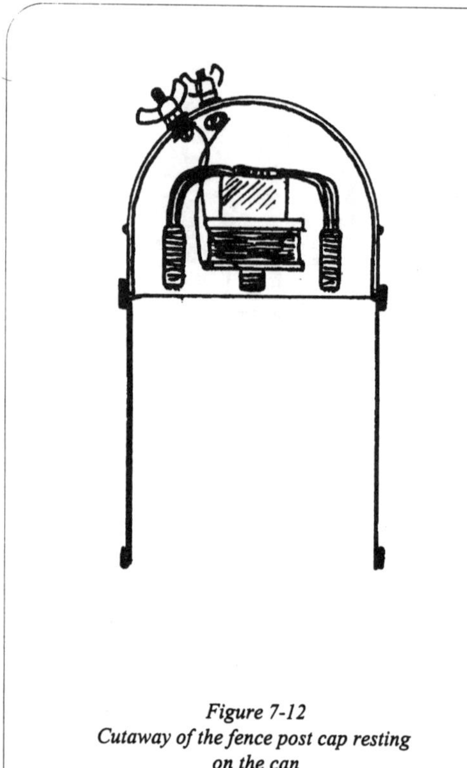

Figure 7-12
Cutaway of the fence post cap resting
on the can

was slipped on to the "W"-shaped core. With the coil and core assembly complete, all that was left to do was come up with a clever way to mount it to the can/diaphragm. (Figure 7-10)

A visit to the hardware store inspired an excellent solution. Fence post caps are dome shaped, and cast from a soft metal alloy, so while they are very rigid and mechanically strong, they are very easy to drill and work. Caps come in varying sizes. My cap was selected because it's diameter allows it to fit just inside of the rim of the can. In this position, it rests on the outer edge of the diaphragm, and is kept from sliding off by the rim of the can.

The cap is anchored to the can with two small metal straps. Figure 7-11 and 7-12 shows the coil/core assembly mounted inside of a fence post cap, and the proper relationship between the cap and the can.

The coil/core assembly is suspended in the center of the dome with a brass bracket. For lack of a better description, the bracket is roughly "U"-shaped, as wide as the inside of the cap, with 1/4" legs. The legs of the "U" are drilled and bolted with 6-32 screws to the walls of the post cap. The coil/core assembly is cemented or epoxied to the bracket. (Figure 7-13)

The trick is to adjust the support bracket so that when the cap is fitted to the can, the poles of the coil/core assembly will be suspended

Figure 7-13
Interior of fence post cap

just above the diaphragm without actually touching it. It may be necessary to elongate the mounting holes in the bracket so that the you can adjust the clearance between the poles and the diaphragm.

The process is a matter of trial and error. Adjust the poles, attach the cap, and check for clearance. You can test for clearance by pushing the diaphragm from the inside of the can (towards the poles). The diaphragm should give just a little and then "click" as it strikes the poles.

6-32 bolts with wing nuts are used as terminals, and mounted on the back of the cap. Since the cap is metallic and conducts electricity, "shoulder" washers are used to insulate the binding posts from the cap itself. Shoulder washers can be salvaged from some electrical junk, or can be home-made from plastic or cardboard. (Figure 7-14)

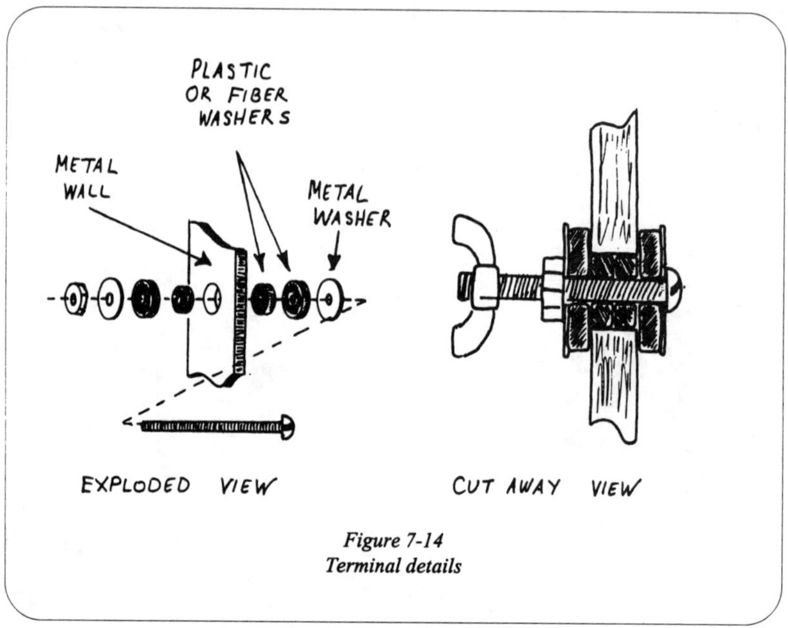

Figure 7-14
Terminal details

End Results and Going Farther

This particular design works exceptionally well. In fact, it compares favorably with the antique earpiece that inspired it, both in tone and sensitivity. It embodies all of the design improvements that I envisioned for the gallows headphone. The core is smaller, and the coil impedance is much higher, thereby reducing's it's tendency to load the receiver. Using it, I have been able to listen to stations across the A.M. band, as well as shortwave broadcasts.

One thing you will notice, however, is that voice and music coming through a tin can tends to *sound* like a tin can. Take an empty tuna can and flick the bottom with your finger. The characteristic "doink" you hear has to do with mechanical resonances inherent in the materials and design of the can. When used as part of a headphone, these resonances can produce overtones which give it an unpleasant metallic sound. Fortunately, this problem is easy to correct.

I discovered that lining the inside of the can with felt "deadens" a lot of the overtones, and diminishes can-induced tonal effects. Ordinary contact cement holds a cylinder of felt to the inside walls of the can. Obviously, the diaphragm is left uncovered.

It is possible to further improve the sound by damping any unusual resonance in the diaphragm itself. By punching a piece of cellophane tape with a paper hole punch, I produced several tiny adhesive disks. Just a few of these disks, when applied to the proper regions on the can's lid, can improved the sound of the headphone without detracting from it's sensitivity. Of course, the number and placement of these disks depends on the sound you are after, and the specific can used in your headphone. This is a trial and error proposition. As they say, your mileage may vary.

The ultimate improvement, of course, is to build a second unit, identical to the first. With a bit of tinkering, it wouldn't be hard to come up with a strap or bridge to couple the headphones, anchor them to your head, pressed to your ears. A headset frees up your hands to work with the receiver itself. Because both of your ears are shielded from external noise, and both are "driven" with audio, the apparent sensitivity is vastly improved.

Chapter 8
The "Cigarette Lighter"
Headphone

In the first chapter of this book, I introduced the idea of analyzing design problems in terms of "essential characteristics" in order to devise solutions. After a discussion of some basic theory, we determined that the essential characteristics of a good headphone included a certain type of coil, certain core materials, a magnet, and a certain type of diaphragm.

Then, I showed you the design for a device employing these principles. Based on the strengths and weaknesses of that first model, we refined our analysis, and designed a second headphone that performs even better. Would it surprise you to find that our analysis was incomplete?

If you take the "essential characteristics" view it's extreme, you would really define a speaker or headphone as nothing more than a "black box" that converts an electrical signal (the audio signal) into mechanical motion (the vibration of the diaphragm). Yes, it *might* include a coil or magnet to move a diaphragm, but nothing says that it has to! In fact, if you could find some other mechanism to perform the electrical to mechanical conversion, you could build a speaker that contained no magnets at all!

Many years ago, it was discovered that certain crystals can produce an electric current when distorted by mechanical stress. This phenomena is known as the *piezoelectric effect*. I've seen laboratory apparatus consisting of a carefully prepared crystal and a large neon bulb. When the crystal is struck with small rubber mallet, it actually produces enough current to flash the neon bulb on and off.

Like many physical processes, the piezoelectric effect is reversible. This means that if we apply an electric current to the crystal, it will mechanically deform itself (it's dimensions will actually change!)

A properly prepared crystal, attached to a diaphragm and stimulated by an electric current, will distort and change shape, thereby causing the diaphragm to move. This is the basis behind the commercially-made piezoelectric earphones found in most crystal radio kits. The question is, how difficult is it to actually build one of these from scratch?

The *Xtal Set Society Newsletter*, volume six, issue three, features a fascinating article by author William Simes. The article describes the complete process of building your own piezoelectric headphone. He begins with a substance called Rochelle salt, from which a crystal is grown. He describes the selection of a suitable crystal, how to properly orient and cut the crystal into strips, and how to assemble those strips into a structure he calls a bimorph. The bimorph crystal assembly is eventually installed in a tin can (sound familiar?) completing the headphone. Don't, however, conclude that the process

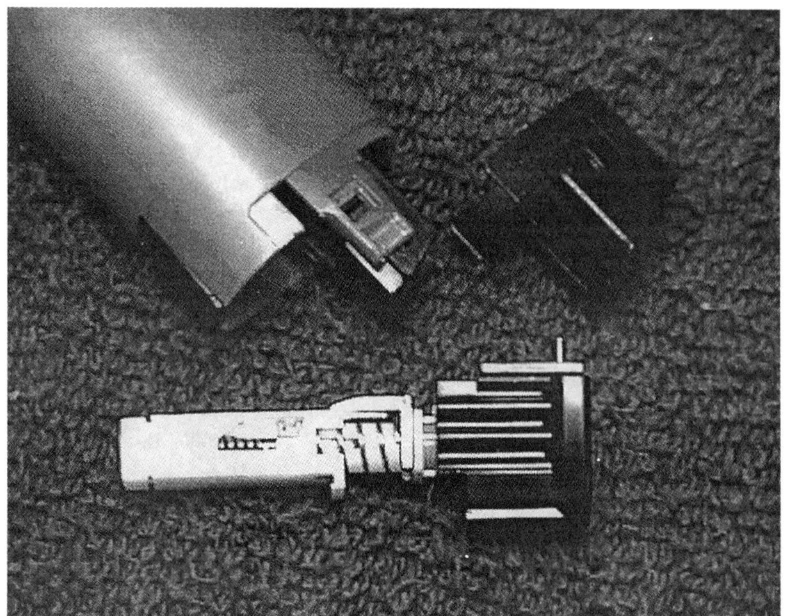

Figure 8-2
Lighter internals

is trivial.

A crystal, by definition, consists of an orderly, repeating arrangement of atoms. These atomic patterns can take the form of very complex, three-dimensional geometric shapes. If you are trying to generate a current by stressing a crystal, the angle and direction of the force with respect to the crystal geometry can make the difference between a large current or none at all. If you want to use the crystal to convert an electric signal to motion, you better know which faces of the crystal to apply the current to. To make a long story short, you can homebrew crystals and build this sort of equipment from scratch, but it takes a lot of patience and a lot of know-how. In my estimation, it's more work than winding a fine coil of wire and fabricating an electromagnetic headphone.

None the less, the article inspired me to think about a simpler design. If I could identify a piezoelectric material in some common household item, and find a way to use it in a headphone design, it would save me all of the effort and trouble associated with crystal fabrication. The solution came to me as I watched a friend light a cigarette. (Figure

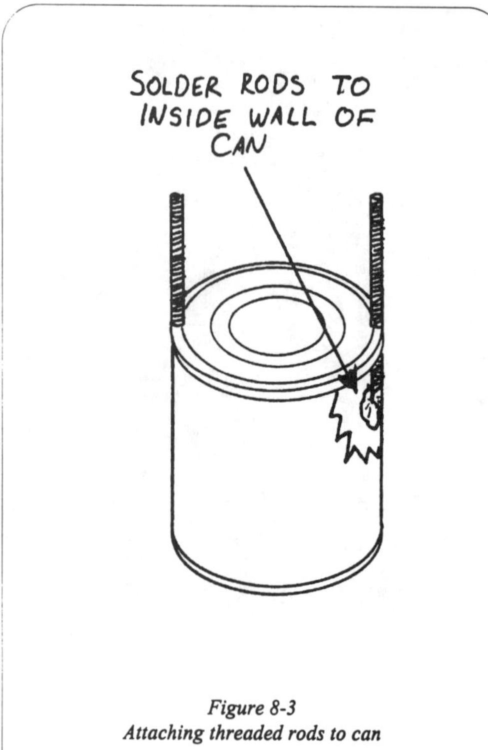

Figure 8-3
Attaching threaded rods to can

8-1)

Most inexpensive, disposable butane lighters use a "flint" wheel to ignite the gas. When spun with the thumb, the rough surface of the wheel grinds against a flint-like material to create a shower of sparks. Less common, but as easily purchased, are the so-called piezoelectric lighters. These use a crystal, like the ones I just described, to generate an electric spark. The spark ignites the butane.

I purchased several piezo lighters at a local gas station and carefully disassembled them. All of them had similar internals. Most lighters of this type break down into two subassemblies. One is composed of the body, fuel cell, and gas jet, and the other is a self-contained spark mechanism. (Figure 8-2)

The spark mechanism features a tiny, cylindrical piezoelectric crystal about the diameter of a pencil lead, and about 1/8" long. The mechanical portion of the spark mechanism consists of a spring and a tiny hammer which is cocked and released to strike the crystal when the lighter's trigger is pressed.

The hammer strikes the bottom face of the crystal, which also serves as an electrical ground. A high voltage pulse appears at the top face of the crystal. This pulse is directed to a pointed metal electrode which is used to ignite the gas.

In short, the cigarette lighter crystal produces electricity when distorted by the spring-driven hammer.

Even though I knew nothing about the composition of the crystal, I knew that it was piezoelectric, and that it generated sizeable voltages in response to mechanical stress. I also knew the axis to which the force must be applied. Given that the piezoelectric effect is reversible, it seemed reasonable to me that if I applied a current to the ends of the crystal, it would probably distort along that same axis. If I linked that crystal to a diaphragm, I should be able to produce sound.

I began construction of the cigarette lighter headphone with a tin can. Like the previous headphone design, the can was selected for it's tone and .007" thickness lid. Two holes were drilled through the bottom of the can, at opposite ends of the lid, and as close to the edge as possible. See figure 8-3. A short length of 6-32 threaded rod was inserted into each hole, and soldered to the inside wall of the can. Note that the rods were *not* soldered to the lid itself.

Next, I fashioned an insulator block, 1/2" by 5/8" in cross section, and about 2 3/4" long. In the device pictured here, the block was made out of a phenolic material, but plastic or wood works just fine. A hole was drilled at each end of the block, spaced so that the block would slide on to the threaded rods protruding from the back of the tin can. The rods and the insulator block form an adjustable suspension mechanism, not unlike the suspension used on the gallows headphone. Washers and wing nuts are threaded onto the rods. (Figure 8-4)

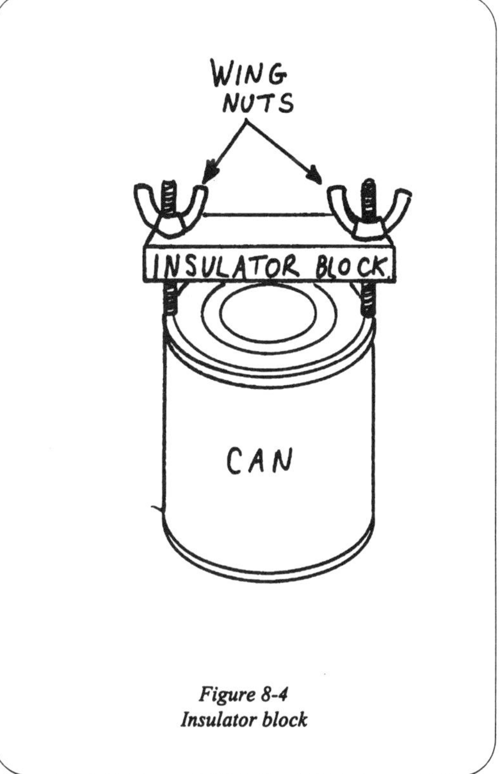

Figure 8-4
Insulator block

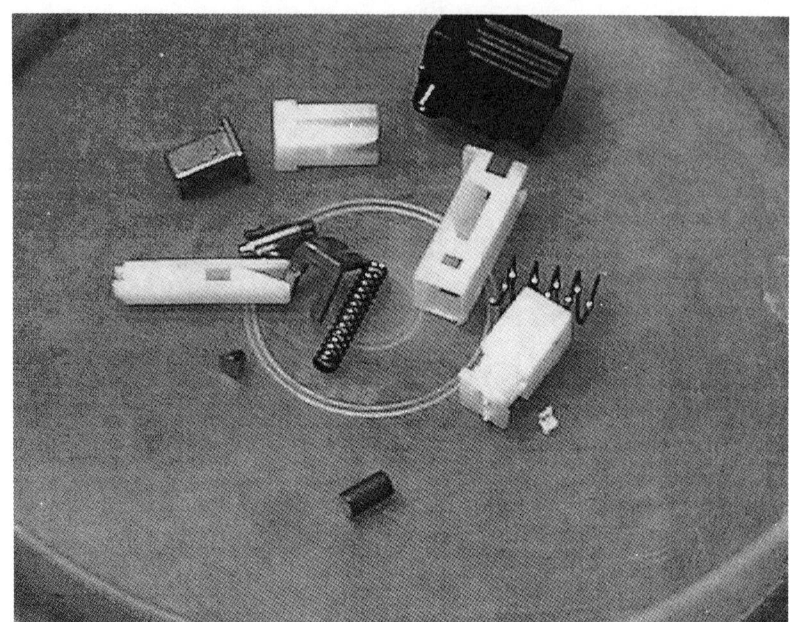

Figure 8-5
Piezo crystal

The next step was to extract the piezo crystal from the lighter. Since there is some variation from one brand of lighter to the next, it's nearly impossible to give you step-by-step instructions. Do exercise patience and *caution*. After all, you are dealing with a device loaded with butane, and butane is *explosive*. Make sure you are wearing safety glasses. If you are not up to the task of reverse-engineering a lighter, consider this chapter academic, and move on.

You will probably find, as I did, that the sparking mechanism can be pulled out of the lighter whole. Throw out the rest of the lighter. Next, carefully break apart the sparking mechanism until you have liberated the crystal. I used a fine pair of wire cutters, a needle, and a razor blade.

In some models, the cylindrical piezoelectric crystal will simply fall out. In other models, the crystal is glued in. The glued crystals are more difficult to remove, but I've done it several times. In any event, proceed with care, as the crystal is fragile and will shatter if treated roughly. Figure 8-5 shows some of the tiny parts in the spark mechanism, and the cylindrical piezoelectric crystal in the foreground.

The final assembly of the cigarette lighter headphone involves the creation of a "sandwich" of sorts. First, I placed a small fiber washer on the center of the diaphragm. On top of the washer, I placed a strip of brass shim stock, about 3/8" wide and 2 inches long, to serve as an electrode. Next, using tweezers, I set the piezo crystal upright, on top of the electrode. A second electrode strip was placed on the top face of the cylindrical crystal. Finally, the insulator block was lowered and the wing nuts were tightened to compress the sandwich and keep it from falling apart. (Figure 8-6)

To make final assembly easier, I cut both electrode strips in advance, and anchored them to the insulator block with 6-32 screws. The end of each electrode was bent into an "L" shape and together they form a holder for the crystal. The crystal is gently pinched between the electrodes' fingers, and stays in place unit the sandwich is properly compressed. (Figure 8-7)

The free ends of the electrodes were terminated with binding posts made from 6-32 bolts and wing nuts. This makes it easy to attach wires to the headphone, and protects the crystal sandwich from being pulled apart by accident.

End Results and Going Farther

The cigarette lighter headphone is a very different animal than the electromagnetic headphones described earlier. Like the other headphones, it has it's share of pros and cons.

On the plus side, this design is extremely simple and quick to build. If you want to build a complete headset (two earpieces), it's a lot easier to whip up two the of the cigarette lighter devices than to build cores and wind coils for the electromagnetic sets. The design is also fairly stable and robust. Time, temperature variations, and an occasional fall to the floor doesn't seem to have changed the device's characteristics.

It's impedance, by the way, is exceptionally high. So high, in fact, that my multi-meter thinks that it's an open circuit (many millions of ohms!). As a result, the cigarette lighter headphone hardly loads crystal receivers at all. Using this headphone, tuning is much sharper and more precise than when the tin can or gallows headphones are used. This is a big plus in urban areas where many stations can be crowded closely together.

On the down side, the cigarette lighter headphone seems to be

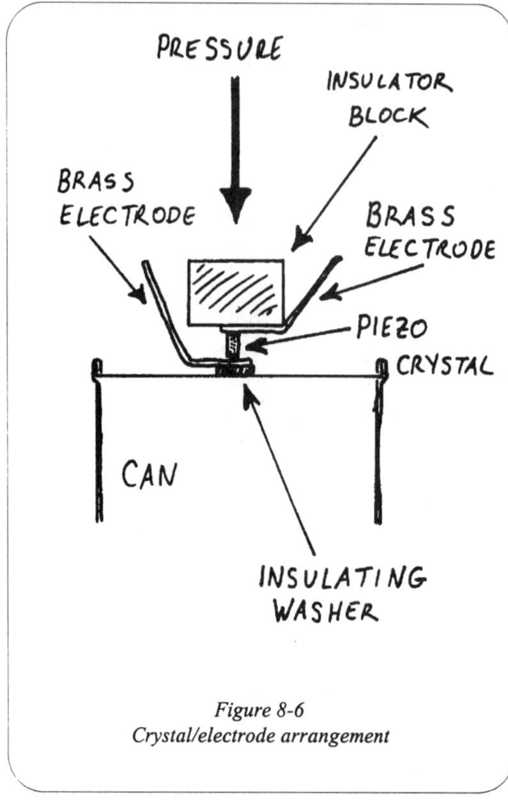

Figure 8-6
Crystal/electrode arrangement

less sensitive than the other designs we've visited. It works best when listening to powerful and/or nearby radio stations. At night, I still prefer the (electromagnetic) tin-can headphone.

Another drawback lies with it's tonal characteristics. The headphone favors high tones over low tones, so the audio it produces always sounds thin and wispy.

Even so, the simplicity of the design, and the ease with which it can be built makes it a good platform for dozens of possible experiments. For example, I've noticed that sound quality and the sensitivity of the device is affected by the adjustment of the wing nuts on the insulator block. The wing nuts can be used to regulate the pressure on the crystal and the electrodes. What is the optimum pressure? Perhaps you'd care to find out.

Another idea involves the use of multiple crystals. Suppose two crystals were stacked, end to end? This might result in a louder headphone if both crystals are oriented properly. (You must be concerned with polarity. You don't want one crystal in the stack expanding in response to a signal while the other one contracts. The net result would be zero! In that case, flip one of the crystals 180 degrees.)

Maybe a better way to go is to use two or more crystals, side by side. Again, polarity is a concern. What if you cut one of the crystal cylinders in half? It's possible that different manufacturers of lighters use different formulations for their crystals. Maybe one brand makes a better headphone than another.

Of course, all of the comments I made about tin can acoustics in the previous chapter apply here as well. Lining the inside of the tin can with felt will surely improve the tone of the headphone. Build two cigarette lighter headphones, one for each ear.

Unfortunately, lousy tone is a characteristic of most p i e z o e l e c t r i c h e a d p h o n e s . However, I've seen radio circuits that feature a resistor connected across the terminals of the p i e z o e l e c t r i c headphone. There are some technical reasons why this may be a good idea, which I won't go into right now. Anyway, resistors are cheap, so try it. Try values between 10,000 and 100,000 ohms.

Happy listening!

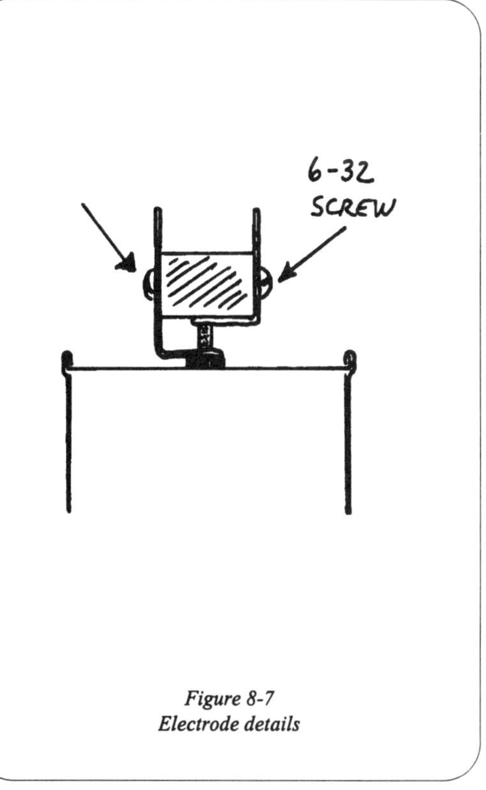

Figure 8-7
Electrode details

Chapter 9
Essential Characteristics Of A Good Detector

I f you recall the comments I made in the chapter on basic theory, I said that a tuner resonates with, or "captures" the signal from a distant transmitter. The tuned signal is passed through a *detector*, which breaks down the transmitted signal into an audio signal. The audio signal, of course, is what's used to drive the headphones.

I also said that the detector accomplishes this through a very selective form of electrical conduction. That is to say, it allows current to pass in one direction, but not in the other. The detector is a type of semiconductor.

Modern detectors are now called *diodes*. Generally speaking, highly purified silicon is purposefully contaminated with certain other materials to yield a "doped" crystal. The crystal Is mounted inside of a glass or plastic bead with wire leads. Production of modern semiconductors occurs at state-of-the-art factories, costing many millions of dollars. The chemistry and mechanics behind the fabrication process is complicated, and excruciatingly precise. Variations in the doped silicon crystal forms the basis for most modern electronics, up to and including desk-top computers.

Exactly how semi-conduction works (on an atomic level) is also beyond the scope of this book. I can tell you that a great deal of scientific and industrial research has gone into that topic, and most of the details are available, if you can stomach higher math and quantum mechanics. Needless to say, duplicating commercial silicon diode processes in the kitchen sink is out of the question.

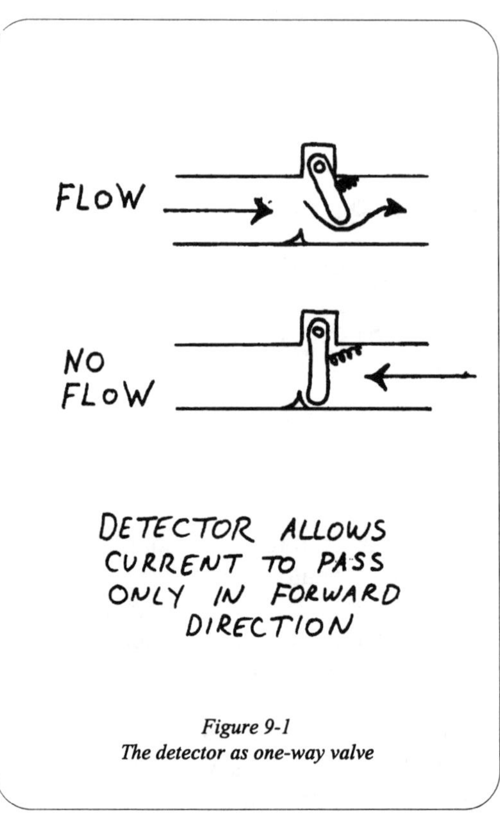

Figure 9-1
The detector as one-way valve

Yet, there are probably thousands of naturally occurring materials that exhibit semiconductor properties, and many can be used to build radio detectors. Materials can be found in the bathroom, kitchen, garage, and sometimes the back yard! In fact, under the right conditions, some of these materials outperform commercial diodes! The trick to selecting home-brew semiconductor materials is to understand the essential characteristics of a good diode/detector.

If a voltage is applied to a diode in the forward direction, the diode allows an electric current to pass through it. If the voltage is applied in the reverse direction, the diode suddenly becomes an insulator, and the electric current is blocked.

In the theory chapter, I compared the detector (diode) to a check valve in a water line. The check valve contains a spring loaded door. When the water flows in the forward direction, water pressure pops the door open, and the water is free to flow on through. If the water tries to flow in the reverse direction, the door slams shut, and the flow of water stops. (Figure 9-1)

What would happen if we modified the spring that closes the door? Obviously, the heavier the spring, the more forward pressure would be required to open the check valve. (Figure 9-2)

Suppose, for the sake of example, that the valve required 10 pounds per square inch to pop it open. Water at any pressure less than

10 psi, when applied to the input, would fail to pass through. For all practical purposes, it may as well not be there at all. Once the input pressure exceeded the 10 psi threshold, however, the valve would begin to open and water would begin to trickle through.

Diodes and crystal detectors exhibit similar behavior. Just like the check valve, detectors will only conduct if the forward voltage exceeds a specific threshold value. This is called the *forward voltage drop*. Different semiconductor materials have

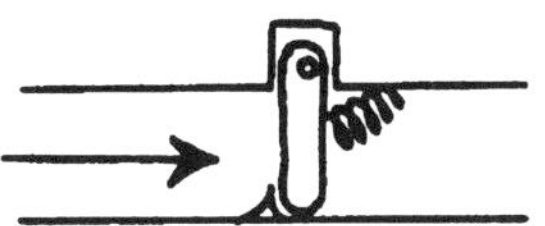

Figure 9-2
One-way valves: heavy vs light springs

different forward voltage drops. Silicon, for example, has a typical forward voltage drop of 0.7 volts. Germanium, on the other hand, has a drop of 0.2 volts.

This has profound implications for crystal radio work. Remember that the crystal radio is a completely passive device. There are no amplifiers, and the electrical signals applied to your detector are only as large as that which your tuner can capture. If the captured signal measures 0.5 volts, and your detector is made of germanium, you should be able to listen to the station. The 0.5 volt signal is greater than germanium's 0.2 volt drop, and should be able to force open the "door" inside of the detector. It would pass through with 0.3 volts to spare.

If your detector was made from silicon, you would probably hear nothing. A 0.5 volt signal is not strong enough to push through the "door" in 0.7 volt silicon. You wouldn't even be aware that the signal existed.

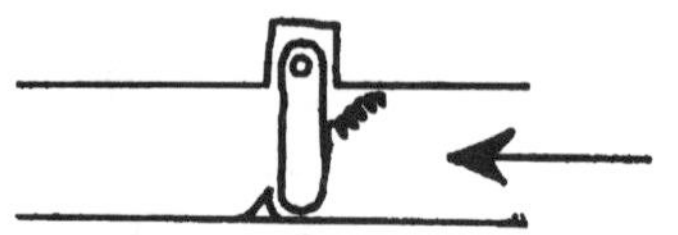

Figure 9-3
Burst valve

The ideal detector would have a *zero* forward voltage drop. It would open up for any forward-flowing signal, regardless how small. Unfortunately, perfect detectors don't exist. For crystal radio work, the goal is to find materials with the lowest forward voltage drop possible. The lower the forward voltage drop, the weaker the signal that you will be able to hear.

Let's look at the check valve again. Suppose we take a small valve and applied water in the reverse direction. The internal door slams shut and blocks the flow. If we begin increasing the pressure, the door will probably seal even tighter. At some point, however, the force of the water will exceed the structural limits of the spring-loaded door. The door will burst or shatter, and the water will blast on through. At that point, the check valve is destroyed, and thereafter, water can flow freely in either direction.

Diodes and detectors behave the same way. If we apply a voltage in the reverse direction, the detector will resist the flow, as long as we don't exceed it's *reverse breakdown voltage*. If that threshold is passed, the detector is destroyed. Current may then flow in either direction, or in some cases, not at all. (Figure 9-3)

Because crystal radio signals are so small, the reverse breakdown voltage of your detector is usually not an issue. On the other hand, you can generate enormous voltages with your own body! Ever get zapped touching a doorknob? It was probably a dry day, and you probably accumulated a charge as you walked across a carpet. The spark that left your hand was many thousands of volts. If you touched your detector instead of the door knob, that spark would probably destroy it. For that matter, it's possible for a very strong local station (high reverse voltages) to degrade or deaden a detector to the point where weak stations can no longer be detected.

When water flowing through a check valve is suddenly reversed, we expect the spring loaded door inside to slam shut and block the flow. The question is, how long does that process take?

Figure 9-4
Closure time

Any door will require a certain, measurable time to close. A big, heavy door will take much longer to close than a small, light door. In either case, water will leak through the valve in the wrong direction until the door has finished closing. If water is sloshed back and forth through the valve very rapidly, you may reach a point where the door is not responsive enough in either direction. It may just wobble in the halfway position, as water shoots through in both ways! (Figure 9-4)

Detectors, like the check valve, will leak briefly when the applied voltage is suddenly reversed. The speed at which the detector recovers from the leak and "seals" properly is called the *recovery time*. The recovery time depends on what materials were used to fabricate the detector.

Radio waves are composed of electrical currents that oscillate. In other words, the voltages flip from forward, to reverse, to forward again, very rapidly. If the recovery time of your detector is too long, it will fail "close" in time to block the reverse flow. This means that it will fail to extract any audio signal that the waves may carry.

Some crystal sets are tuned to listen to shortwave signals. Shortwave signals are interesting to listen to, because they can include morse code and foreign broadcasts. Electrically, they are just like the signals from a local A.M. radio station, except that they are transmitted at a much higher frequency. Be aware that a detector that works fine for A.M. may be useless for shortwave. If you find this to be the case, blame it on the recovery time of your detector.

There is one final variation on the check valve analogy. Up to this point, we've assumed that when the door is open, water flows through with no restriction. When the door is closed, the valve is totally water tight. In reality, there are losses in water pressure any time there is flow through a valve. This might be due to friction or even turbulence. If pressure is applied in reverse and the door seal is worn, it may leak a tiny bit, even when fully closed. In other words, no mechanical device in the real world is absolutely perfect.

For similar reasons, no detector is perfect. It may pose some resistance to currents flowing in the forward direction, and may allow the leakage of some current in reverse. As long as current flow is significantly greater in the forward direction than in reverse, you have a detector. The greater the difference between forward flow and reverse leakage, the better the detector.

How would we describe the a good detector, then? We've covered a lot of ground in the last few pages, so let's summarize the highlights: A good detector allows forward current to flow much more easily than reverse current. A good detector must respond rapidly to reversals in current, and a good detector should have as low a forward voltage drop as possible. This is to allow it to work on very faint signals.

Chapter 10
Practical Detector Designs

N ow that we have a handle on what the detector does, and what constitutes a good one, let's look at a few examples from the real world.

Figure 10-1 shows a diagram of one of the earliest successful detectors. Structurally it was simple. It consisted of a small cup of nitric acid. A tiny platinum wire just touched the surface of the acid. One terminal of the detector was platinum, and the other terminal was the cup containing the acid. When properly adjusted and used, it was considered the finest detector of it's day.

This is an example of the so-called *electrolytic detector*. The detector's behavior stems from an effect that occurs at the junction between the platinum and the acid.

You will find that this is a recurring theme. Often, when dissimilar conductors (or semiconductors) are brought together, the point of contact forms a detector. While I'd recommend against playing with nitric acid, there are probably other common fluids that would exhibit similar behavior. I've never tried to replicate this ancient design.

Another interesting form of detector is fabricated from a razorblade, a pencil lead, and a safety pin. (Figure 10-2) The design is classic: The razorblade is screwed flat to a small block of wood. The safety pin is opened and a screw is driven through it's head into the wood. This forms an inverted "V" whose point is suspended over the razorblade. The pencil lead is attached to the point of the safety pin with a little bit of wire. The safety pin is adjusted again so that the lead just touches the side of the razorblade. The safety pin becomes one terminal, and the razorblade the other.

Razor blades are often coated with special materials,

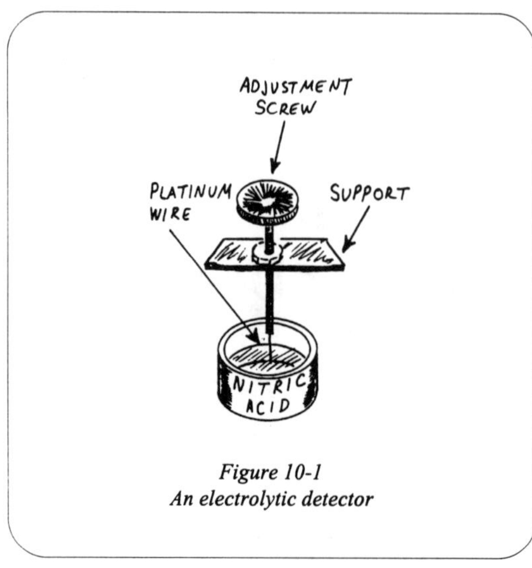

Figure 10-1
An electrolytic detector

presumably to make them resist corrosion. Pencil leads are not lead, but carbon. The detector, in this case, is formed by the gentle contact of the carbon against the blade's coating.

Detectors of this type require some tinkering. The safety pin may need adjustment to apply more or less pressure to the carbon. The point of contact may have to be moved. Some areas of the blade will be more "sensitive" than others.

There is a great opportunity for experimentation here. What if you used something besides pencil lead for the contact point? What if you applied the safety pin's point directly to the blade? Try it. Try using something other than the razor blade, too. A steel washer? A penny? Tinfoil? I've heard rusty blades specified as being better than clean ones.

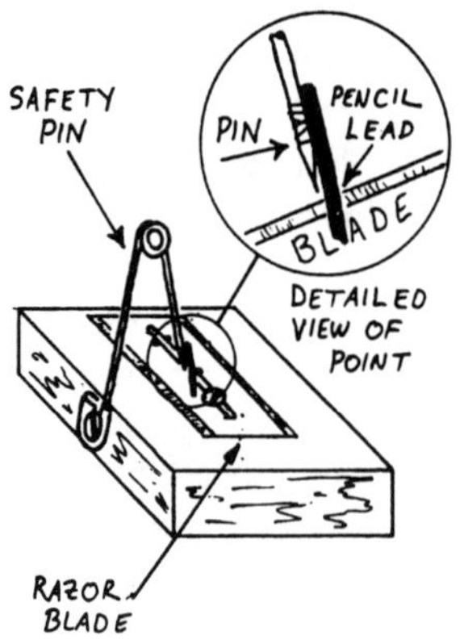

It's hard to tell what will work until you actually try it. Remember- there are several characteristics that make a good detector, and it's a matter of give and take. Materials that are mediocre in one respect may be excellent in another. There's no reason why you can't build

Figure 10-2
A razor blade and pencil lead

different kinds of detectors suited for different purposes. One detector may be best for strong local stations. Another may be best for high frequencies or distant night time reception. I have seen dozens of variations on this basic design.

Detector action, by the way, can sometimes be demonstrated using an ohmmeter. If you measure a detector's

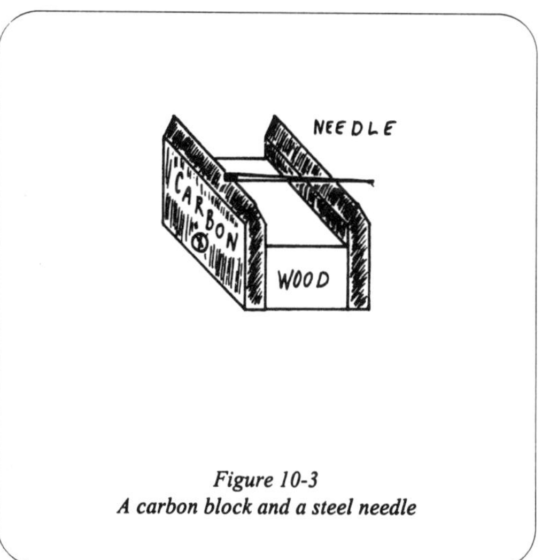

Figure 10-3
A carbon block and a steel needle

resistance in one direction, and then reverse the terminals and measure again, you should see a difference.

Another variation on the carbon/steel detector is shown in Figure 10-3. Two pieces of carbon are each filed to a taper, like the cutting edge of a chisel. The carbons are attached to a small block of wood, so that their sharpened edges point upward. A common sewing needle is placed on the edges so that it crosses from one carbon to the other.

If there is no sign of detector action, bump the needle with your finger. Again, some contact spots are more sensitive than others. Carbon, by the way, can still be salvaged from the

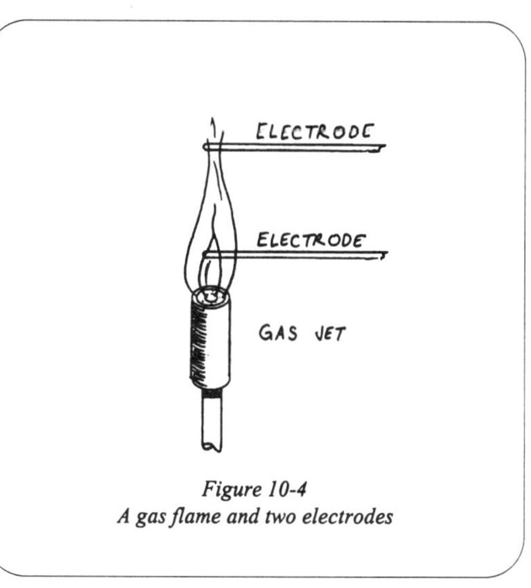

Figure 10-4
A gas flame and two electrodes

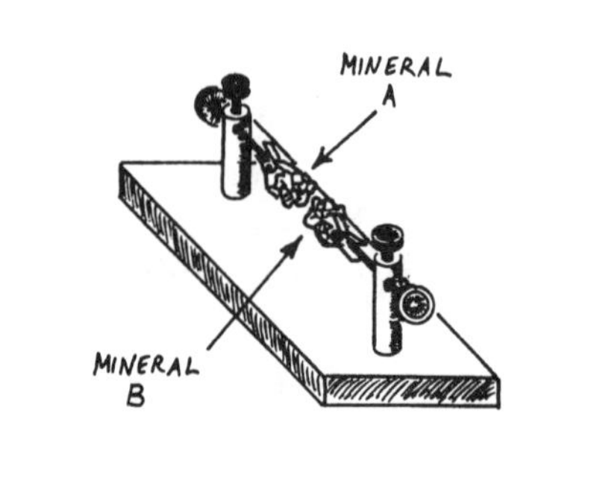

Figure 10-5
A pair of different minerals in contact

core of old carbon/zinc flashlight batteries. Don't, however, try to take apart alkaline or nickel/cadmium batteries. They contain nasty chemicals.

You may find carbon/steel detectors to be insensitive to weak signals. Their sensitivity can be vastly improved through a technique known as *biasing*. Biasing is a technique for compensating for the high forward voltage drop of some detectors. I'll describe the process in greater detail later.

One of the most interesting detectors I've ever read about appeared in a back issue of the Xtal Set Society's newsletter. (Volume 7, Number 2) Author Nile Steiner described the insertion of two horizontal metal electrodes into a gas flame. The scheme is said to produce a workable detector. (Figure 10-4)

Each electrode exploits the characteristics of a different portion of the flame. Current, the author reported, would readily flow from one electrode to the other, but not as well in reverse. To be honest, I tried it, and was unable to observe the same effect. Perhaps my flame was too hot, or not hot enough. I may have used the wrong materials, or spaced the electrodes improperly. In any event, it's an intriguing idea, and worth more experimentation.

Last, but certainly not least, is the so-called "crystal" detector. It represents the technology that gives the crystal radio it's very name. Best of all, it is exactly what it sounds like — radio detection through the use of metallic crystals.

Few metals occur in pure form in their natural state. Most metals are more or less chemically reactive, and are found in nature only in combination with other elements. Often, these mineral compounds are partially conductive. When placed in contact with other materials, they can produce exceptional detectors.

I have some very old books with engravings of popular crystal detectors of the day. Figure 10-5 is drawing of one such detector. Two battery clamps, each mounted to an adjustable arm, were attached to a wooden base. In the mouth of each clamp was a chunk of some metallic mineral. The clamps were then adjusted so that the two mineral samples were brought into contact. The point of contact between the two samples forms the detector.

One of the oldest and best of the crystal detectors was

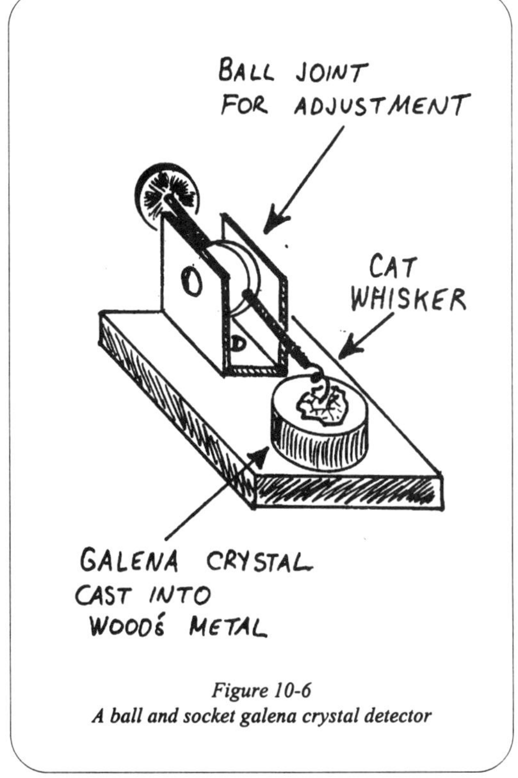

Figure 10-6
A ball and socket galena crystal detector

based on a mineral called *galena*. Galena is a shiny, grey-colored mineral composed of lead and sulfur. If probed with a tiny wire, called a "cat's whisker" the point of contact between the galena and the wire forms a very sensitive detector. One terminal of the detector is the galena crystal itself, and the other terminal is the cat's whisker.

Physically, crystal detectors are delicate devices. The action of the junction between two detector materials depends upon the physical point of contact, and the pressure with which contact is made. What this means is that if any old piece of wire is randomly poked against any old galena crystal, odds are that the resulting detector will not function properly. You generally have to probe a galena crystal until a "sweet spot" is found.

Building a practical detector requires the creation of a mechanism that will support the crystal and the whisker, yet allow the point of contact to be changed at will. The pressure with which the

Figure 10-7
A vintage galena crystal detector

whisker contacts the crystal must also be adjustable. Finally, the detector mechanism must be stable enough so that it will steadfastly maintain those adjustments until intentionally changed.

Figure 10-6 shows a drawing of a common galena detector design, and Figure 10-7 shows a picture of an actual vintage device. Because crystals are difficult to attach wires to, it was once common to cast the crystal into a small metal button. The button, in turn was snapped into a metal cup or socket, specifically designed to accept it. The crystal was embedded in the metal, but one of its faces was always left exposed. This was the face that would later be probed by the whisker.

Like modern semiconductor materials, galena can be damaged by excessive heat. So, instead of casting the crystal into a metal like lead, the old timers used an alloy called "Wood's metal". Wood's metal looks like lead, but has a very low melting point. As a matter of fact, Wood's metal will actually melt in hot water!

The genuine detector features a short metal rod. The rod is fitted with a insulated handle at one end, and the cat's whisker at the other. The rod passes through a metal ball, which in turn is clipped into the mouth of a "U"-shaped bracket. This arrangement has the effect of creating a ball-and-socket joint. The whisker end of the rod can easily be adjusted to the left, right, up, or down. The rod is also free to slide in and out of the ball, to provide back-and-forth adjustability.

By taking advantage of the ball and slide properties of the rod, the whisker could be placed anywhere on the face of the crystal. Once adjusted, the detector would continue to work for a long time. The ball and socket approach was very popular, and detectors of this design came in many different shapes and sizes.

Chapter 11
The "Boom" Detector

n the previous chapters, I outlined essential characteristics of a good detector, and then went on to describe some real-world examples. To summarize, the performance of any crystal detector depends upon the materials, the point of contact, and the pressure at that contact. In this chapter, I'd like to show you how these requirements were translated into piece of equipment that is not only functional and practical, but attractive as well. (Figure 11-1)

The galena ball-and-socket detector, described earlier, is a wonderful design. It allows for full adjustability, and is easy to work with. My first instinct was to try to duplicate that design with common materials. The problem turned out to be the ball itself. If one owns a lathe, turning a brass or aluminum ball is probably not an overly difficult task. I didn't own a lathe and wouldn't use it if I had, having limited myself only to hand tools.

Digging through a junk box, I found a round brass ball that had been removed from the center of a lever-type kitchen faucet. Unfortunately, the ball was riddled with holes and grooves to channel the flow of water. I tried to fill the channels with solder, and discovered that, whatever that ball was made of, it wasn't brass. Solder refused to stick to it.

Next, I noticed that the fire screen in front of our fireplace was capped with decorative brass balls, about 1/2" in diameter. Unfortunately, on closer inspection, they turned out to be some sort of pot metal, plated to look like brass. Besides, they were hollow. I eventually gave up on the ball-and-socket design, deciding that there must be some other way.

Once again, inspiration came from an unexpected source. I own

Figure 11-1
The "boom" detector

some nice microphones which are used for recording music. The simplest mike stands are nothing more than a cast iron base, and a vertical tube. The mike, of course, is then fastened to the top of the tube.

Better stands feature a "boom". The boom consists of pivoted metal block that is threaded to the top of the mike stand, and a second tube that slides through the block. The mike is then attached to the end of the boom. The boom can be pivoted up into the air to suspend the mike above a piano, or downward to mike the bottom of a drum. The boom can be slid in and out of its block, so the sound source can be either close to, or far from the mike stand. The effect is that the mike can be supported anywhere within a two or three foot radius of the stand. (Figure 11-2)

I envisioned the mike stand and it's boom scaled down, with the microphone replaced by a cat's whisker. The full flexibility of the boom would satisfy the adjustability requirement of a practical detector mechanism.

The boom was fabricated from a piece of brass rod. Each end

was threaded with a 6-32 die. On the back end, I attached on old plastic knob which forms the handle. The other was fitted with two 6-32 nuts. Using the finest bit that I could find, I drilled a tiny hole through the rod in the threaded area between the nuts. To attach the cat's whisker to the boom, you feed it through the tiny hole, and tighten the 6-32 nuts on either side. I'll

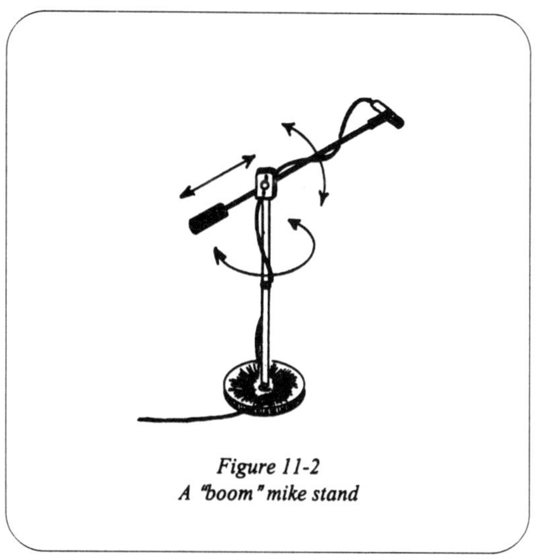

Figure 11-2
A "boom" mike stand

discuss the whisker itself in just a moment. (Figure 11-3)

The boom stand is an "L" shaped piece of heavy brass stock, 2 1/4" tall with a 1 1/8" base. The "L" was located on the detector's wooden base, and a hole was drilled through both the "L" and the wood. Next, a piece of thin brass stock was cut to the shape of a frying pan. The round portion of the pan is about 1 3/4" in diameter, and the handle is 1/2" wide and 3/4" long. A hole is drilled in the center of the pan. The "frying pan" is inserted between the wooden base and the brass "L". (Figure 11-4)

A 6-32 bolt passes upward through the wooden base, through the frying pan, through the base of the "L", and through a brass

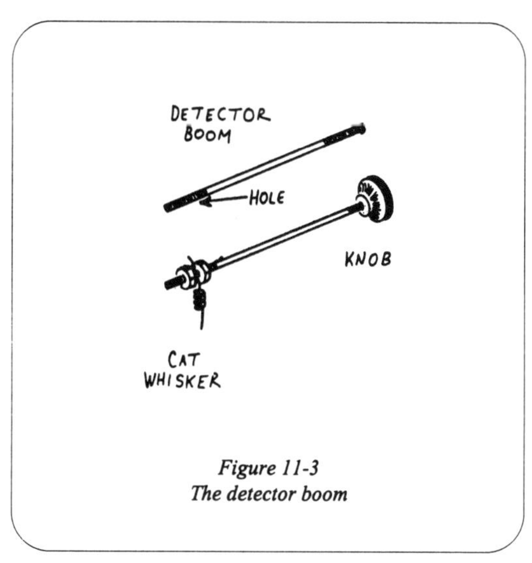

Figure 11-3
The detector boom

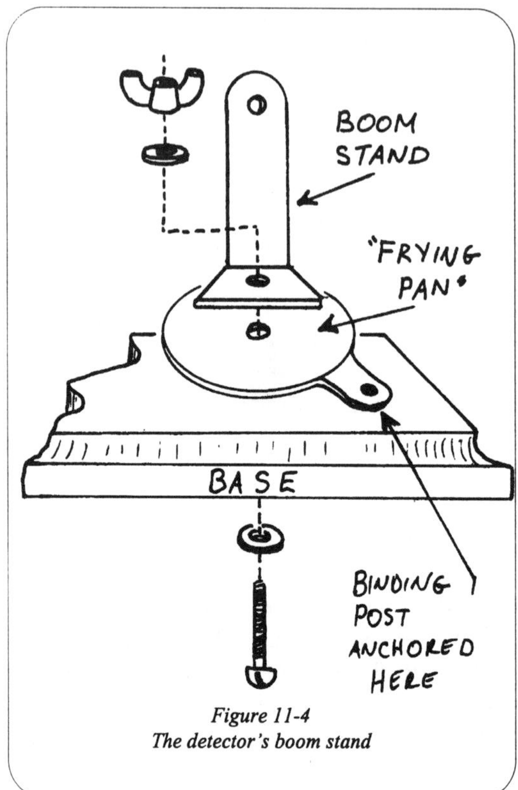

Figure 11-4
The detector's boom stand

washer. It is terminated with a wing nut. For the record, the brass washer was made by drilling a hole through an old brass video arcade token. A penny would work, just as well.

This arrangement allows the "L" to pivot left to right. When the wing nut is tightened, the "L" is locked into place. The "frying pan" beneath the "L" serves two purposes. First, it provides a smooth surface for the "L" to rotate upon, and it makes excellent electrical contract with the "L". A hole was drilled through the handle of the pan and through the wooden base. A 6-32 bolt and some brass nuts create a binding post on which to connect wires.

The boom slide bracket was fashioned from a strip of brass stock, about 5/16" wide. The strip was bent into a "U" shape, so that the curve of the "U" has a radius of about 1/8". A hole was drilled from one leg of the "U" to the other, about 1/8" from the curve. The hole is the same diameter as the brass rod used to fabricate the boom.

Next, using a hacksaw, I cut a slot through the base of the "U" to both holes. The boom rod was slid through the holes. A 6-32 bolt was inserted through the space bounded by the bottom of the "U" and the side of the boom rod (the bolt is inserted perpendicular to the boom rod). The bolt is terminated with a small washer and a wingnut. (Figure 11-5)

When the wingnut is loose, the boom rod can slide back and forth through the bracket with little effort. When the wing nut is tightened, the bolt tends to pinch the hacksawed slot closed, which reduces the diameter of the slide hole. The boom is then locked into position.

The boom pivot is nothing more than a brass game token, with a hole drilled through it's center. The free ends of the slide bracket were bent outward at right angles (like little wings), and soldered to the game token. A hole was drilled through the top leg of the "L", and the boom pivot was bolted to the "L". A 6-32 bolt passes through the boom pivot, through

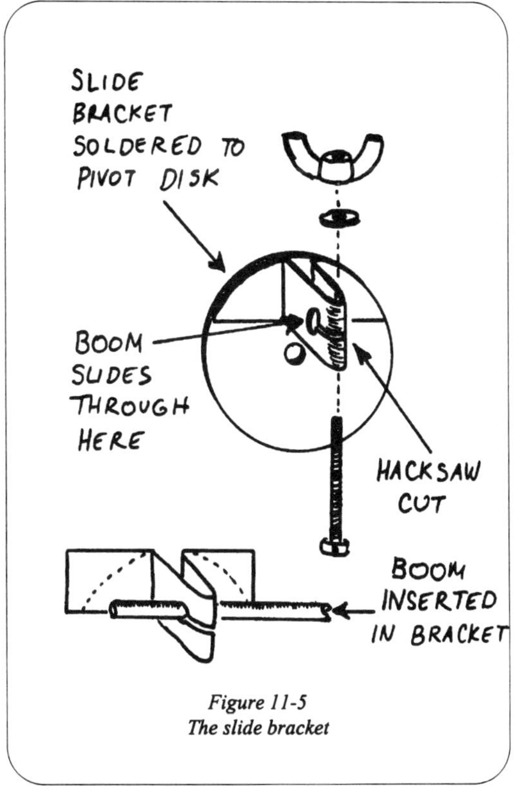

Figure 11-5
The slide bracket

the "L", and through a brass washer. It's capped with a wing nut.

When the boom pivot wing nut is loose, the pivot allows the boom to be raised and lowered. When the wing nut is tightened, the boom is locked into position. (Figure 11-6)

That completed the boom assembly. The base pivot allows the boom to swing from side to side. The boom pivot allows the boom to swing up and down. The boom slide allows the boom to slide in and out. The boom is fully adjustable, and can be locked in any position when the nuts are tightened.

Next, my thoughts turned to the crystal and it's mounting. I wanted to build a galena detector, but I wasn't certain that it would work for me. Besides, I wanted the option of being able to try more than one mineral. Even if I had settled on a mineral at that point, I couldn't employ a traditional mounting. I had no Wood's metal, and I knew that casting a crystal in molten lead might end up destroying it. I needed a novel

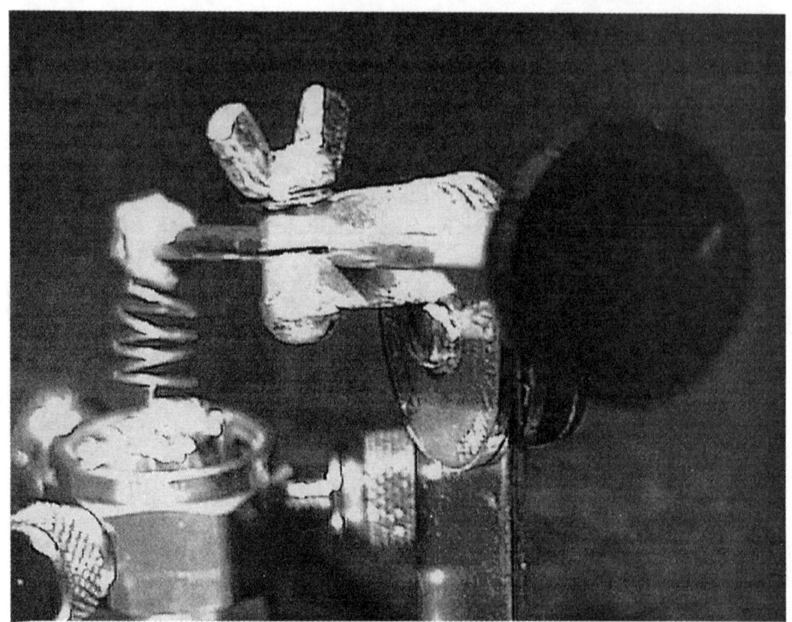

Figure 11-6
Slide bracket details

crystal holder that was simple, adaptable, and would allow an easy change of crystals.

My solution appears in Figure 11-7. My crystal holder was fabricated from a 3/4" gas fitting. Through it's side, I carefully drilled three equally spaced holes. The holes were tapped for 6-32 threads. I found three nice thumbscrews, salvaged from who-knows-what, and threaded them into the holes. You could use ordinary machine screws if you wanted to. These screws will clamp and secure the crystal sample.

Some brass scrap was soldered to the bottom of the fitting to form a floor. A hole was drilled through the floor and through the wooden base. A 6-32 bolt secured the crystal holder to the detector's wooden base.

An oblong strip of thin brass stock was cut, and a hole was drilled near each end. One end of the strip was sandwiched between the wood and the crystal cup. The other end of the strip was terminated with a 6-32 bolt and some nuts to form a binding post. The purpose of the strip is to provide an electrical connection to the crystal cup.

Figure 11-7
The crystal cup

To load the crystal cup, I began with a piece of aluminum foil. The foil was folded over and over to form a thick strip, about 3/8" in width. The strip was then wrapped around the waist of my galena crystal sample. Next the crystal, with it's aluminum belt, was placed inside of the crystal cup. The thumbscrews were tightened, as equally as possible, until the screws had bitten into the foil, and the crystal was locked into place.

The soft foil provides protection against crushing the crystal, because it absorbs and distributes some of the pressure and stress. Also, because the foil conforms to the shape of the crystal, it ensures good electrical contact. (Figure 11-8)

The final piece to this puzzle is the cat's whisker. My whisker was made from the "B" string of an acoustic guitar. The wire was wrapped four times around a pencil, to form a loose spring. The ends of the wire were trimmed and bent to point outward from the ends of the coil, parallel to it's axis. One end was inserted into the tiny hole drilled into the end of the boom, and locked into place with two 6-32 nuts. The other end of the coil was placed in contact with the face of the galena

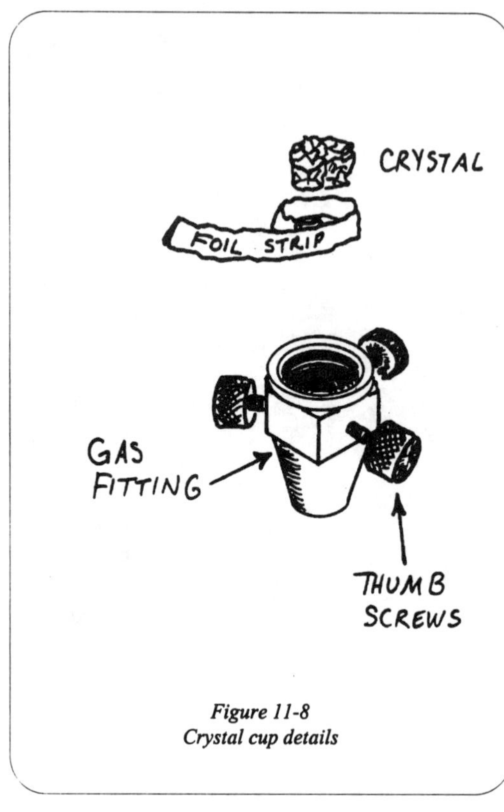

Figure 11-8
Crystal cup details

crystal.

The wooden base is another example of the ready-made wooden plaques available at any craft store. The detector was built on 3" by 5" pine. As is my practice throughout this book, all holes were drilled and all parts were fitted to the wooden base while it was in it's unfinished state. When the detector was complete and all the bugs were worked out, I dismantled it, and finished the wood. After treating the wood with stain and a coat of polyurethane, I reassembled the detector. Green felt was cemented to the bottom of the detector as a finishing touch.

End Results and Going Farther

This detector design, as described, has worked *very* well. I have tried a variety of crystal set circuits using this detector, and have been pleased in all instances. In fact, you can receive and listen to radio broadcasts with nothing more than the boom detector and one of the headphones I've already described.

On the other hand, if you've gathered anything from my preceding comments, you'll conclude that success with old-time detectors is as much a matter of intuition and art as science. I played with several materials and techniques until I arrived at a usable configuration.

For example, I began with a cat's whisker made of brass wire. For me, it did not work at all. I tried several different wire materials, and finally settled on the guitar string. It seemed to work well, electrically. Mechanically, it has some spring to it, which allows me to control the pressure it applies to the crystal. There's no end to the wire materials that could be tested. I've thought about using the wire taken from a ball-point pen spring. Phosphor bronze wire is recommended in vintage literature.

I built my first detector using galena. As I mentioned earlier, galena is universally regarded as one of the most sensitive detector materials. I live in Tucson, Arizona, which yearly hosts one of the biggest gem and mineral shows in the world. There, I found a beautiful sample of galena from a mine back east somewhere. The problem was, that it made a lousy detector! I concluded that not all galena is alike, and the difference between a good galena detector and a bad one may have to do with which specific mine it came from! In retrospect, it makes sense. Different impurities in the crystal can either enhance or degrade it's electrical behavior.

To make a long story short, I tried all sorts of metallic minerals in this detector, including purified silicon. The silicon was purchased as scrap from an outfit that makes modern, commercial semiconductors. The strangest material I tried was taken from a grinding machine. As a grinding wheel bites into metal, it shears off tiny particles of metal. The shearing process creates so much heat, that the particles glow, creating a shower of sparks. In this case, the glowing particles thrown by the wheel collected in a corner of the grinder, and welded themselves together to form a gray, crystalline stalagmite. A machinist saved a chunk of this material for me, and suggested that I try it. You'd be surprised what will work in a pinch.

I finally settled on a fragment of iron pyrite, more commonly known as "fool's gold". The crystal has proven to be stable, and sensitive. "Fool's gold" is available at any rock shop, and is extremely cheap. I've even seen it pawned off at garage sales, as part of "mineral collections", that were initially purchased in a tourist gift shop or souvenir stand.

Adjustment of the detector is a matter of trial and error. If the detector is connected as part of a radio circuit, the headphone will produce noticeable crackling as the whisker is moved across the face of the crystal. At some point or another, detection will begin, and you will hear music or speech. Sometimes a change in whisker pressure will dramatically improve the station's volume, other times the detector will quit working all together. At that point, you need to relocate the whisker and try again.

One final note concerning the adjustment of the guitar string cat's whisker: Jabbing the point of the whisker directly onto the face of the crystal often works. However, the most sensitive settings I've found occurred when the *side* of the whisker, not it's point, was in contact with the crystal. Can you offer a theory why?

Chapter 12
Essential Characteristics Of
The Condenser

The various headphones and detectors described in the preceding chapters are sufficient to build a device capable of receiving and decoding radio signals (I'll show you a specific circuit in a later chapter). While this "bare bones" approach is interesting, it suffers from two major flaws. First, without equipment for tuning, you're likely to receive more than one station at a time. The second problem is sensitivity. If your receiver is not properly tuned, it'll do a poor job of capturing the energy from the radio waves you're interested in. So, even if you hear something, you'll find that the signal isn't nearly as strong as it could be. A tuner is a must.

In the chapter on basic radio theory, I tried to explain the transfer of energy from a radio transmitter to a receiver using tuning forks as an example. I also stated that the electronic equivalent of tuning fork can be fashioned from a combination of two electronic components. In this chapter I'd like to introduce you to the first of these, namely, the condenser.

Imagine taking a small, square sheet of window glass, measuring 6" on a side. Next, imagine cutting two squares of aluminum foil, each measuring 5", and gluing one aluminum square on each side of the glass. As simple as this device may sound, it exhibits some interesting properties. (Figure 12-1)

If the foil plates are temporarily connected to a source of electricity, which is then removed, you'll discover that the plates retain a charge. If one plate is then connected to the other with a piece of wire, current will flow until the charge is equalized. This basic ability to store

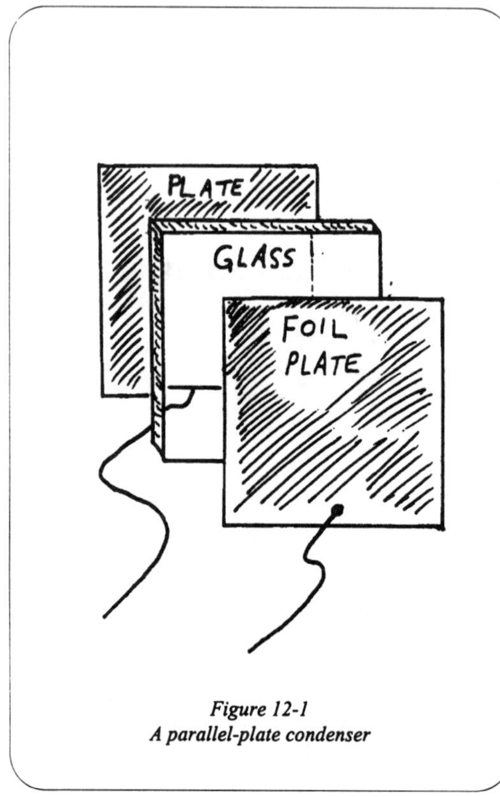

Figure 12-1
A parallel-plate condenser

an electric charge is called capacitance. Modern books call this device a *capacitor*. The old timers called the same thing a *condenser*.

In some ways, applying a charge to a condenser is like inflating a balloon. You have to exert effort to force your breath into the balloon. The energy you expend doesn't vanish, but is stored in the form of the compressed air and the tension in the walls of the balloon. When the balloon is discharged, by releasing the neck, the stored energy is released as rapidly moving air. The balloon zips away at high speed, bouncing this way and that until the stored energy has been depleted.

All conductors exhibit a certain amount of resistance to the flow of electricity. For practical purposes, that resistance is fixed, and doesn't change with normal use. Condensers, on the other hand, are picky about the currents that flow through them. Condensers easily pass currents that oscillate (change back-and-forth). In fact, the higher the frequency, the easier they flow through. As frequency drops, however, the currents find it more and more difficult to pass through. Direct current, which does not oscillate, is completely blocked. This prejudicial behavior against low frequencies is called *capacitive reactance*.

"Low frequency" is a relative term. The frequency at which a signal can no longer pass through the condenser depends on capacitance. A condenser with a great deal of capacitance will pass lower frequencies than a condenser with very little capacitance.

$$C = \epsilon \frac{A}{d}$$

Figure 12-2
The formula for parallel plate capacitance

The fact that the condenser is a frequency sensitive device should suggest to you that it has promise as a tuning device.

Figure 12-2 shows the basic formula for capacitance. Notice that capacitance is directly related to the area of the foil plates. If you want to build a condenser that has more capacitance, you simply make the foil plates larger. Capacitance is inversely related to the distance between the plates. If you use a thicker piece of glass, the distance between the foil plates will increase, so the capacitance will actually go down.

What's the funny symbol in the equation? That's the Greek letter epsilon. Epsilon is a physical constant that is determined by the material used to separate the plates. If you use a fixed set of plates set apart at a fixed distance, the capacitance of your condenser will depend one what material fills the space between them. A condenser with glass between its plates will have five to ten times the capacitance of an identical condenser gapped with air. Engineers call the insulating material between the condenser plates the *dielectric*.

Capacitance is measured in units called *farads*. One farad is an awful lot of capacitance, and there are few applications for condensers this large. By comparison, most radio applications condensers are a million or billion times smaller. For convenience, engineers created secondary units like the microfarad or picofarad. A one microfarad condenser measures one one-millionth of a farad. A one-picofarad condenser measures one one-millionth of a millionth of a farad.

Notice that this formula doesn't specify foil, glass, or square shapes. It doesn't care about those details. All that is important is the size of the conductive plates, the distance between them, and the dielectric material between them. Taken to it's extreme, you can say that *anywhere* you separate two conductors with an insulator, you form a condenser!

The choice of plate and dielectric materials can alter the

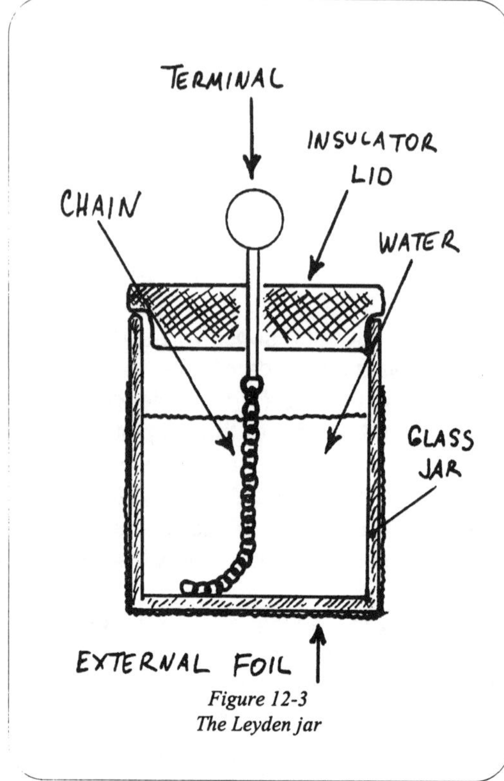

Figure 12-3
The Leyden jar

performance of condensers in ways other than by affecting capacitance. Remember the balloon analogy? We stored energy by forcing gas, under pressure, into the balloon. When we released the neck of the balloon, we got all of the energy back. Suppose, however, that the balloon has a pinhole. If we inflate it, pinch the neck, and simply wait, the air will eventually leak out. The gas that escapes through the hole is considered a loss, because any energy lost through the pin hole can't be recovered at the neck of the balloon. In fact, if you wait long enough, the balloon will go completely flat. All of the energy you expended inflating the balloon will have been lost.

In the real world, all condensers exhibit similar losses through a variety of mechanisms. If we store energy in the condenser, and then try to withdraw it, we never get all of it back. A certain percentage is always lost. The trick is to select your materials so that these losses are minimized. Copper, brass or aluminum makes for decent plates. Glass, plastic, waxed paper, or air usually makes a suitable dielectric. For the home made condenser builder, experimentation and an open mind is the key to success.

For the record, the glass and foil condenser is not something I created simply for the purpose of example. Plate glass condensers have been used for years in experimental, high power radio signal sources like Tesla coils. Because glass is an excellent insulator, it is possible to apply very high voltages to the foil plates, without fear that a spark will punch through.

I recently restored an ignition coil from a Model T era automobile. Inside, I found a small version of the glass and foil condenser, just as I've described. In the case of the ignition coil, the condenser helped served two purposes. It helped lengthen the life of the electrical contacts that switched the coil on and off, and at the same time, helped the coil produce more energetic sparks.

Glass and foil condensers are easy to make, and easy to modify. They offer excellent opportunities for crystal radio experimentation.

Glass for picture frames is cheap and readily available at any hardware store. What can I say about aluminum foil? It's everywhere. Foil can be stuck to the glass with glue, wax, or even varnish. When used as part of experimental high-voltage equipment, the condensers are sometimes dipped in wax, or immersed in oil.

Don't assume that condensers have to be flat. One of the oldest practical condensers, called a Leyden jar, is cylindrical. A typical Leyden jar can be fabricated from a clean, smooth-sided glass jar. The outside of the jar is wrapped in foil, to half of it's height. The inside of the jar is filled with water to the halfway mark, and a few drops of acid are added to the water. The jar is covered with a wooden lid. A small brass chain is suspended from the wood lid, and lies partially immersed in the water. (Figure 12-3)

"This is a condenser?" you ask. Absolutely. The foil wrapping acts as one plate. The wall of the glass jar acts as the dielectric. The water inside the jar is conductive, particularly after acid has been added it. It behaves like a second conductive plate. The chain provides a convenient way to "connect" to water. The capacitance of the Leyden jar depends on the area of the foil plate, and the type of glass used to fabricate the jar.

The Leyden jar was created in the early 1700's, nearly simultaneously, by two independent experimenters. Some credit a Dutch physicist Musschenbroek, while others credit a German by the name of von Kleist..

Large Leyden jars can store enormous charges, large enough to be lethal. They were a staple item in early laboratories studying electricity.

For crystal radio work, jars of this size are overkill. Small Leyden jars, on the other hand, may well be appropriate. In any event, they illustrate that workable condensers come in a rainbow of shapes, sizes, and materials.

For instance, two sheets of brass separated by a piece of waxed paper makes a useable condenser. Two pieces aluminum, cut from beer cans, can be separated by a piece of cardboard. Two metal pie plates, insulated from each other by an old record forms a condenser as well. As long as basic condenser principles are embraced, the lists of plate materials, dielectric materials, and condenser geometries are almost limitless.

In the earlier glass plate example, I said that increasing the size of the plates increases the capacitance. Bigger plates will require a bigger dielectric in order to keep the plates separated. So, as the capacitance is raised, the physical size of the condenser must become larger and larger. You can see there comes a point when further increases in physical size become ridiculous. Maybe you can only purchase glass up to a certain size, or foil up to a certain width. Or, maybe you just don't have the space for a six foot condenser! It stands to reason that any factor that limits the physical size of the condenser must eventually limit it's capacitance, right? The short answer is yes.

The long answer is that there is a way around a limitation of this sort. If you need a 10 microfarad condenser but can't build one because of it's size, you can instead build two 5 microfarad units and combine them! In fact, any number of condensers can be linked together.

To do this, you connect the condensers in parallel. All of the left hand plates of all of the condensers are connected together. This forms one terminal. All of the right hand plates of all of the condensers are connected together. This forms the other terminal. The result is a "super" condenser that has the capacitance of all of the contributors combined. It acts as though it had large plates with an area equal to the sum total of the plate areas of each contributor.

Connecting condensers in a chain, or in series, will have an opposite effect. The super condenser that results will end up having a combined capacitance that is *less* than the smallest capacitance value in the chain.

Now that we've discussed the essential characteristics of the condenser, let's summarize them before we move on.

First, whenever two conductors are brought near each other and are separated by an insulator (the dielectric), they create a condenser. Condensers have a property called capacitance, which allows then to store electric charges. It also allows condensers to discriminate between electrical currents with different frequencies.

The capacitance of a condenser depends on the area of the

internal conductors, the distance between them, and the dielectric material between them. Finally, if two or more condensers are connected together in parallel, they create a "super" condenser whose capacitance value is the sum total of all of the contributors.

Chapter 13
The "Paper Tube" Condenser

Of all the home made radio parts described in this book, condensers are the easiest to fabricate. There really isn't all that much to them. Just the same, one of my stated goals from the very beginning, was to produce "lab-style" components to experiment with.

One of advantages of using commercial condensers (cheating!) is that these devices are small, uniform, durable, convenient, and stable. I wanted to devise a process for building homemade condensers with as many of these advantages as possible. Let's take a look at what I came up with. (Figure 13-1)

I began construction of my condenser with the selection of plate and dielectric material. I was fortunate enough to stumble upon some copper foil. Copper foil is available in craft stores for embossing. It is also sold by some electronics dealers as a material with which to shield sensitive electronics. Since copper is a better conductor than aluminum, I used it because I could get it. Note that aluminum foil would have been very nearly as good.

Most office supply stores sell clear plastic sheets for use on overhead projectors. The sheets measure 8 ½ by 11 inches, the same as regular paper. They can be loaded into any laser printer, to produce transparencies for business presentations. This plastic is thin, flexible, and reasonably resistant to heat. This is the material that I choose as my first condenser's dielectric.

I fabricated the first plate from a strip of copper foil, 1 1/2" wide and 12" long. Notice that the strip was cut with a 1/4" tail, about 1" long. Figure 13-2 shows how the tail was folded once, and then twice, so that it juts out perpendicular to the body of the strip. The finished plate

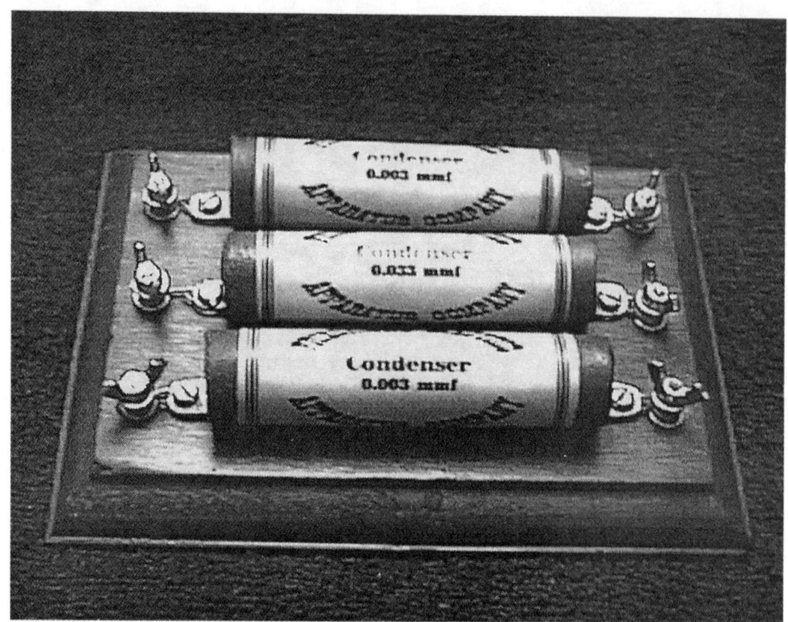

Figure 13-1
"Paper tube" condensers

assembly looks somewhat like the letter "L". I then repeated the above process to produce a second plate.

The dielectric strip was cut from the plastic, 2" wide, and about an inch longer than the plate strips. For reasons that will become obvious in just a second, I produced two identical dielectric strips.

To assemble the condenser, I set one of the dielectric strips on the table, running left to right. On top of it, I set the first plate. The plate was centered vertically and horizontally on the dielectric, and oriented so that the "tail" of the plate pointed downward from the left end of the plate.

I set the second dielectric strip on top, followed by the second plate. This time, the plate was oriented so that it's "tail" pointed upward from the left end of the plate. Again, the plate was centered.

Starting from the end with the tails, I slowly wound this plastic and foil sandwich into a tight roll. I used a pencil as form on which to begin winding. When the roll was complete, I kept it from unraveling with a piece of cellophane tape, and removed the pencil. What I was left with was a small cylinder, with a foil tail dangling from each end. These were the terminals for the condenser. (Figure 13-3)

Why go to the trouble of building a condenser this way? The first advantage should be obvious. Rolling your plates into a cylinder reduces plates with large areas into compact cylinders. The second advantage is takes a bit of explaining.

In the case of flat condensers, like the foil and glass unit

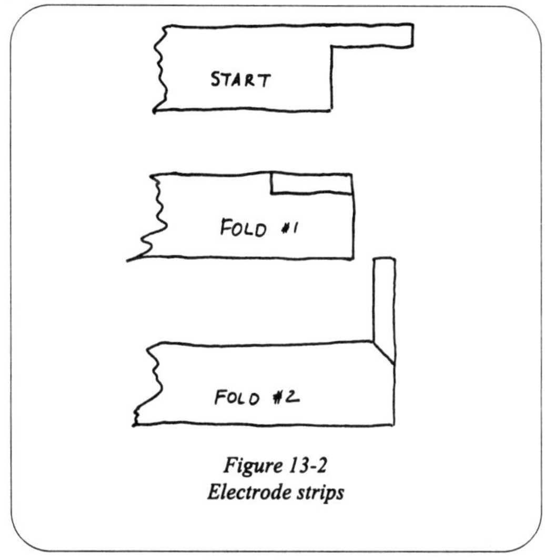

Figure 13-2
Electrode strips

discussed earlier, only the sides of the plates that are facing each other contribute to the device's capacitance. The outer sides of the plates face away from the dielectric, and really don't contribute anything to the process. As far as capacitance is concerned, the area of the plates facing outward is wasted space.

Rolled condensers are different. If you study the roll end-wise you will notice that the plates alternate as you move from one layer to the next. Put another way, each plate is surrounded on both sides by a dielectric and the *other* plate. Because each plate interacts with the other on two sides instead of one, it's almost like having twice as much foil with

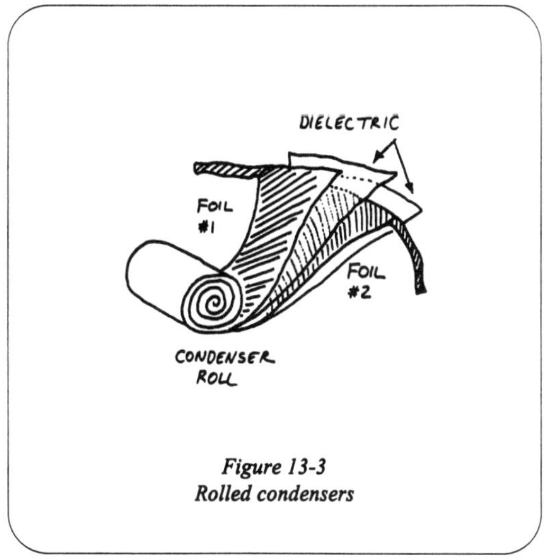

Figure 13-3
Rolled condensers

Figure 13-4
Rolled condensers

twice the area! As a result, the capacitance approaches double the value it would be if the plates and dielectric were laid flat on a table top.

Note that the improvement is always something less than double, because the last few inches of one of the plates ends up comprising the outer surface of the cylinder. The area of the plate facing outward from the skin of the roll can't contribute to the bonus.

Now that the heart of the condenser was complete, I wanted to create a sturdy package for it. (Figure 13-4) Always on the lookout for salvageable materials, I found a heavy cardboard tube, about 7/8" in diameter and 8 ½" long. The tube was the discarded core of a roll of fax machine paper. Using a very sharp knife, I cut off a 4" length. This would be the body of my condenser.

From thin brass stock, I cut a strip about 3/8" in width. Setting the tube up on end, I used a sharp knife, to cut a slit in the wall of the tube. The slit was 3/8" wide, and parallel to the end of the tube. It was located about 1/4" from the end. I inserted the brass strip into the slot, and folded it over with pliers to form a terminal. Figure 13-5 shows how this was done. The process was repeated on the other end of the tube to create a second terminal.

Next, the condenser roll was slipped inside of the tube. The delicate foil "tails" coming from each end of the roll were soldered to the brass terminals mounted on the tube. (Figure 13-6)

At this point, I stopped to make sure the condenser had not been damaged. I checked it with my ohmmeter. Good capacitors of this type will always show infinite resistance. If the plates, through rough handling, somehow short together, the condenser is no good. The ohmmeter will show the low resistance caused by the short. In my case, the condenser was just fine.

The next step was to seal the condenser in wax. I had a large chunk of raw beeswax from another project, and decided to use it as my sealing compound. I set a shallow pan of water on the stove and brought it to a boil. In the water, I place a clean, empty tin can, and in the can I placed my wax. This "double boiler" arrangement quickly melted the wax without danger of overheating the wax and catching it on fire.

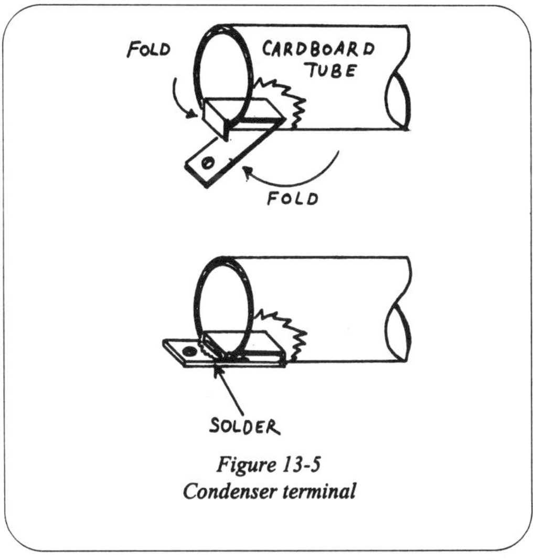

Figure 13-5
Condenser terminal

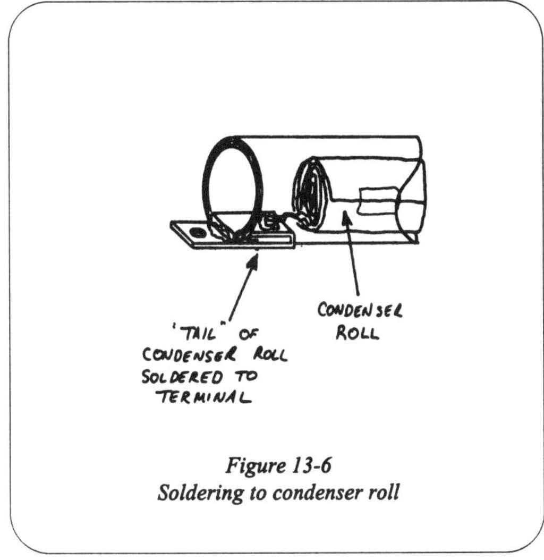

Figure 13-6
Soldering to condenser roll

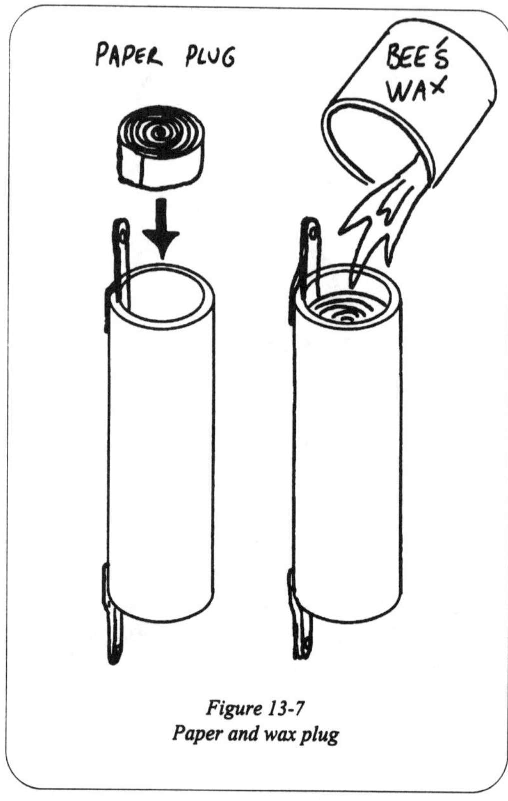

Figure 13-7
Paper and wax plug

Never melt wax through the direct application of heat.

The filling process actually took several steps. First, I produced a paper plug by winding a narrow strip of scrap paper into a cylinder. The plug was inserted into the left side of the condenser, and pushed in about 3/8". I turned the condenser left-side up, and poured wax over the paper plug to produce a nice cap. I set the condenser in the freezer to set. (Figure 13-7)

When the left cap had hardened, I flipped the tube so that the right side was up, and began filling the entire condenser with wax. I filled the tube to withing about 1/4 inch of the top, and tapped the sides of the tube with a pencil to encourage any trapped air to bubble to the surface. Again, I placed the condenser in the freezer.

Finally, when the interior wax had set, I added wax to the open end to produce a nice cap.

At this point, the device was finished. I took the artistic liberty to produce some vintage-style paper labels and pasted them to the body of the tube. The entire assembly was given a liberal coat of polyurethane.

For convenience, I built a condenser bank composed of three such condensers, mounted on a nice pine base. The base, of course, was stained and sealed with polyurethane. I used 6-32 machine screws and wing nuts as terminals.

End Results and Going Farther

The first condenser I built using the above process was connected to an LCR bridge at a local laboratory. An LCR bridge is a very precise way of measuring condensers. It reported a value of 0.003 microfarads. I went home and built two more condensers using the same technique, and later had those measured, too. All three were in extremely close agreement. This is important, because it means that the process is very repeatable. Once you arrive at the proper plate sizes, the process will produce condensers with very predictable values.

The fact that the condensers have been sealed in wax, as well as urethane, has made them very stable. After more than six months of seasonal changes in temperature, humidity, and air pressure, I have detected no significant change in capacitance.

While I submit that this technique produces good-quality home-build condensers for homebrew crystal radios, there is ample room for experimentation.

For example, I think that good condensers could be made using aluminum foil instead of the copper foil that I used. The only trick would be to devise some way to attach the foil tails from the condenser roll, to the brass condenser terminals. Maybe a crimp or clamping arrangement would do the trick.

As far as dielectrics are concerned, there are a lot of options. I built one condenser with 1 and ½ " by 12" plates, insulated with the heavy plastic film which a catalog came sealed in. Tightly rolled, it resulted in a capacitance of 0.002 microfarads.

I also built several condensers using waxed paper. A wax paper condenser with the same plates as above, measured 0.006 microfarads. Wax paper appears to have a larger epsilon value than various plastics that I've tried, so for a given set of plates, it will produce condensers with more capacitance.

I think that plastic from a grocery or garbage bag would be interesting, as would facial tissue. Another idea would be to try a piece of nylon fabric.

Another tip: When you roll the condenser, try to keep track of which tail connects to the plate that ends up on the outside of the roll. Mark the body of the condenser to denote this somehow. Later, if the condenser is used in some circuit where one terminal is grounded, make sure that the marked tail goes to ground. By doing this, the condenser will be self-shielding, and less likely to cause you tuning problems when you bring your hands near it.

I mentioned using beeswax to seal the tube. I like beeswax, because it seems more firm than candle wax, and less inclined to get soft from the warmth of your hand. Ordinary candle wax is probably easier to find, and I'm sure it would do just as well. Don't forget that grocery stores sell blocks of clean, white paraffin wax for canning purposes.

Whether you ultimately build the paper tube condenser or simple flat plate condensers, I'm sure you'll eventually want to build condensers with other values. I've already introduced the capacitance formula to you, and in theory, you can use it to get a rough idea of how your tube condenser will act before you actually build it. The problem will be determining what value of epsilon to use.

Physics books sometimes list epsilon values for common materials. You'll find values for glass, wax, rubber, and paper. Of course, unless the glass, wax, rubber, or paper you have is the same type as was measured in the lab, the epsilon values will at best be approximate. You can forget finding meaningful values to apply to plastics. So, how do you design for a specific capacitance?

The no-equipment approach is to first build a whole bunch of condensers with varying plate sizes. When it comes time to building up radio circuits, you can use them on a trial-and-error basis. Or, you can design the condensers based on an air dielectric, and then multiply the value you compute by a "fudge factor" determined through experimentation.

If you have access to an LCR bridge, or a handheld meter that has a capacitance function, you can build up a few devices, measure them, and then make some educated guesses about what plate sizes to use for smaller or larger condensers.

A more scientific approach, I suppose, would be to build a "standard" condenser. Take two square plates of some known size, put a dielectric in between them, and then measure the capacitor you created. If you work the equation backwards, you can compute the epsilon value for the dielectric you used. Once you know the epsilon, can then design any size condenser with that dielectric, without further need of special test equipment.

Chapter 14
Practical Variable Condensers

All of the condensers described so far are designed to produce a specific capacity. You might say that these are "fixed" condensers because, once assembled, they produce a single, unchanging value of capacitance. If a different capacitance becomes necessary, the only option is to build and install a different condenser.

One way to minimize the inconvenience of having to physically install and remove condensers from a circuit is to build a substitution box. A substitution box consists of a bank of different condensers, wired to a selector switch. Rotating the switch selects a specific condenser from the bank. The drawback to this technique is that your capacitance will always be limited to the values in the bank. In other words, no in-between values can be generated.

A better solution would be to create a condenser that is adjustable *by design*. Conceptually, this is rather simple. If you recall, a condenser's capacitance depends upon the area of the plates, the dielectric material between them, and the distance between them. If any of these parameters are varied, the capacitance will vary with it. In other words, the essential characteristic of a variable condenser is that it features a mechanism to vary area, distance, or dielectric.

Let's start with the dielectric. As you would expect, two plates separated by a fixed distance forms a condenser. The dielectric in this case is air. Now imagine taking a sheet of plastic, and inserting it slowly between the plates. As the sheet advances, it's dielectric qualities will begin to have a greater and greater affect on the total capacitance. As the air is forced out of the gap, it's dielectric contribution will be proportionately reduced. The condenser becomes variable, because it's

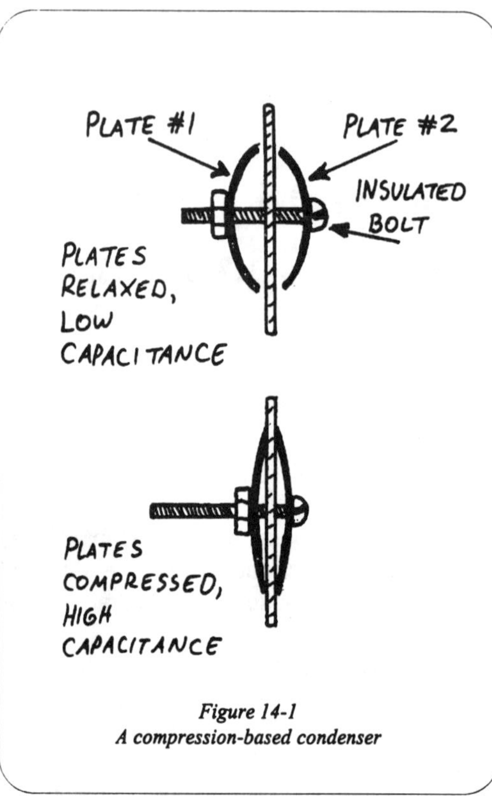

Figure 14-1
A compression-based condenser

capacitance depends on the extent to which to the plastic is inserted.

While the idea is theoretically sound, I don't know of any commercial examples in radio work. The device as just described is conceptual. My guess would be that the technique does not produce enough of a span between it's high and low capacitance values.

Manipulating the distance between the plates to control capacitance seems like a better approach. Figure 14-1 shows a diagram of a variable condenser design that was once fairly common. Basically, it's built just like the flat plate condenser from previous chapters. The difference is that each plate and the dielectric has been drilled , and an insulated screw passes through the entire sandwich. The plate material is selected to be thin and somewhat springy, with a tendency to bow or curl away from the dielectric.

This increases the effective distance between the plates, and the capacitance value is reduced. When the screw is tightened, this has the effect of compressing the sandwich and bringing the plates into closer proximity. By tightening the screw, the capacitance can be raised. Ultimately, capacitance depends on the degree to which the insulated screw has been tightened.

Observe that when a hard cover book is closed, the pages are compressed, and are confined to close proximity with each other. As the book is slowly opened, the pages tend to fan out, and separate themselves from each other. If each page was a dielectric, surrounded on either side by plate material, it would create a variable condenser whose capacitance could be adjusted by opening and closing the book. Years ago, a friend of mine who collected antique radios told me that this very idea had been used in some very early radio sets.

Without doubt, the most common way to vary capacitance is to vary plate area. In principle, this involves two overlapping plates, one of which is stationary, and the other, which is moveable. A dielectric, of course, lies between the plates. Since capacitance only occurs where the plates overlap, the capacitance can be changed at will by simply adjusting the extent of the overlap.

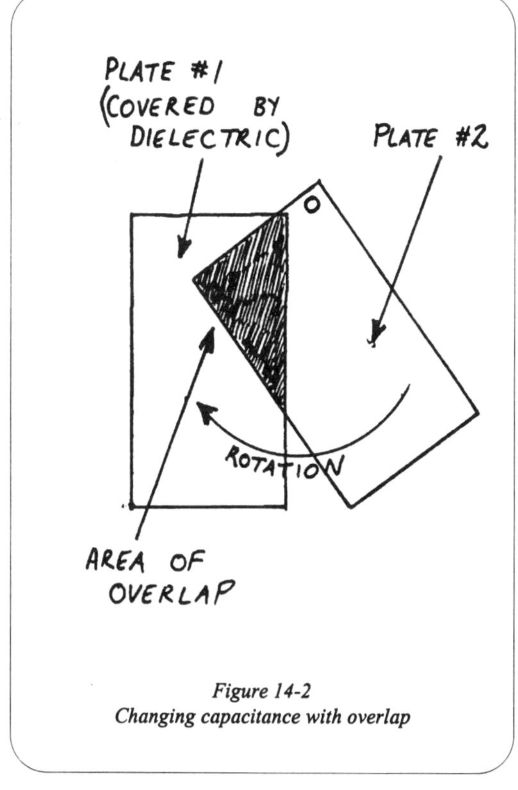

Figure 14-2
Changing capacitance with overlap

Strictly speaking, we don't really change the area of the plates themselves. What we do is control the effective area of the condenser, by controlling the amount of plate area that actually contributes to capacitance.

Figure 14-2 shows a physical implementation of the idea. One plate, covered with a suitable dielectric material is attached to a wooden base. The second plate is drilled at one corner, a screw is inserted, and the plate is screwed to the block. The screw is left just loose enough to allow the top plate to pivot on the screw. The top plate is located so that, as it rotates, it overlaps the bottom plate to a lesser or greater degree. Rotating the top plate on it's pivot causes the effective (overlapping)

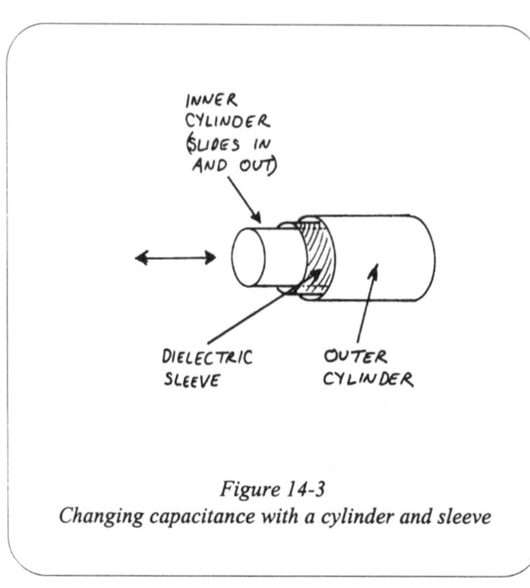

Figure 14-3
Changing capacitance with a cylinder and sleeve

area to change, which causes a corresponding change in capacitance.

One design I like a lot consists of two metal tubes or cylinders. The smaller of the two is coated or wrapped in a dielectric material, and sized so that it just slides into the larger cylinder. The cylinders, of course, act as the condenser's plates. Sliding the smaller tube in and out of the larger tube changes the amount of overlapping plate area, which results in a change in capacitance. (Figure 14-3)

Here's another approach: Imagine a long, flat rectangle of metal, fastened to a wooden base. This would comprise one plate of the condenser. The plate, of course, is covered with a dielectric layer. The top plate is not thick and flat, but is actually a metal film or foil, wound around a wooden shaft. When the top the plate is fully wound, with the roll resting on the dielectric, the capacitance is very small. As the shaft is rotated, the top plate begins to unwind and lay flat, like a carpet that's being unrolled. The more foil that is unwound, the more area there is to react

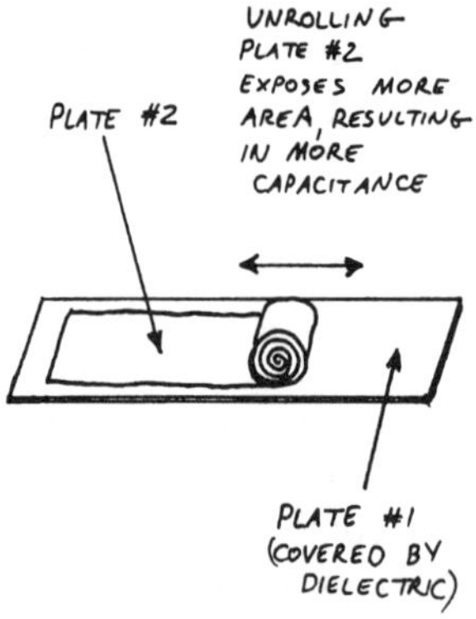

Figure 14-4
Changing capacitance with rolled-up plate

with the fixed plate, and the greater the capacitance will be. (Figure 14-4)

An excellent implementation of this idea appears in the book *Crystal Set Projects*, published by the Xtal Set Society. Experimenter Eric Hudson describes a very clean design based on what he calls a "Rolamite bearing". His materials include a few small blocks of wood, tinfoil, packaging tape (the dielectric), and some PVC pipe for rollers.

Even the primitive Leyden jar condenser can be made variable! Draining water from the Leyden jar reduces the water level. This causes a reduction in effective (internal) plate area, and a corresponding reduction in capacitance. Adding water will increase the Leyden jar's capacitance until the water level exceeds the height of the foil on the outside of the jar.

Experimenter William Simes, also appearing in *Crystal Set Projects*, refined and modernized this idea with a two-liter plastic soda bottle. Two equal-sized rectangles of foil were glued, side by side, to the outside of the bottle. When water was added to the bottle, each plate formed a condenser with the water. Since both the condensers shared the water as a common plate, they acted like a single condenser. (Figure 14-5)

A wooden cradle was fashioned to support the bottle horizontally. The bottle was not filled full of water. Only enough water was placed in the bottle to completely cover the side facing downward.

Capacitance varies like this: When both foil plates are on the down-side of the bottle, they are completely "covered" by the water on the inside of the bottle. This represents the maximum overlap between the foil plates and the water. The relative capacitance in this position is high. As the bottle is rotated, the plates begin to rotate out from under the water. As the area of foil occluded by water is reduced, so is the capacitance.

Water, in and of itself, is a poor conductor. So, Simes added salt to the water to make a saturated solution. In another case, he used water with a small amount of muriatic acid (hydrochloric acid). Kerosene was then added to the acid solution to prevent it from "wetting" the inside of the bottle. The kerosene prevents the water from sticking to sides of the bottle as it is rotated, and sharpens the boundary between the wet and dry areas.

Without doubt, the most common variable condenser design was the rotary type. Probably all commercial radios, from the 1920's to the late 70's feature at least one of these. The rotary condenser consists of two parts: a fixed plate, called a stator, and a moveable plate

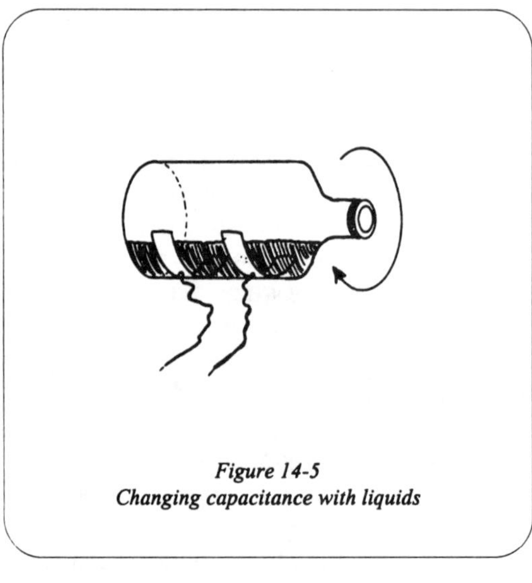

Figure 14-5
Changing capacitance with liquids

called a rotor. (Figure 14-6)

In it's simplest form, the rotor plate is shaped like a half-circle, and mounted on a shaft. The shaft is supported by bearings of some type. As the shaft is rotated, the rotor plate comes into proximity of the stator plate, and the amount of overlap between the two plates determines the capacitance. The plates, of course, never actually touch. They are separated by a very small gap, meaning that air is the dielectric.

If you recall, I told you that two or more condensers can be combined for more capacitance, by connecting their plates together in parallel. Most commercial implementations of the rotary condenser exploit this by using multiple rotor and stator plates. As the rotor shaft is turned, the rotors plates mesh with the fixed stator plates. (Figure 14-7)

Perhaps you're giving some thought to building one of these yourself. If you use a large stator and a large rotor, and limit yourself to two or three plates, it's probably not too big a task. Building a

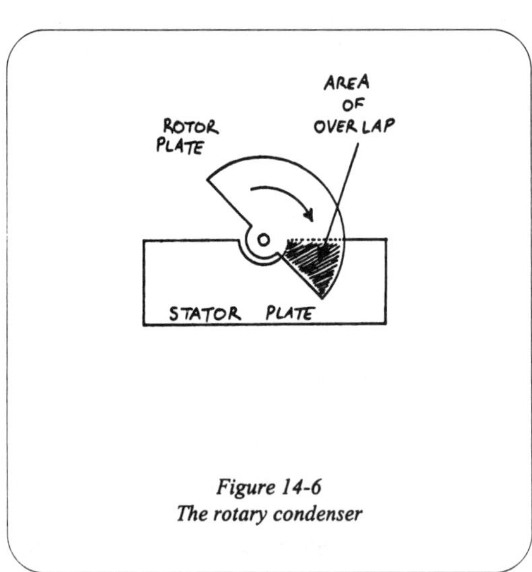

Figure 14-6
The rotary condenser

condenser with multiple rotor plates is another matter entirely.

The challenge lies in the precision with which the components must be fabricated and assembled. All of the plates, both rotor and stator, must be perfectly parallel. In addition, the rotor plates must be p e r f e c t l y perpendicular to the shaft. If so much as a single plate is skewed or bent, even a tiny bit, it will touch an adjacent plate. This creates a short circuit, that makes the condenser unusable.

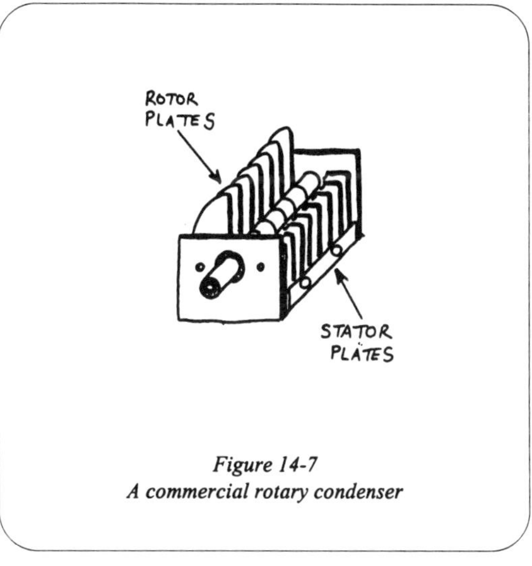

Figure 14-7
A commercial rotary condenser

I cut a series of rotor plates from brass, stacked them with cardboard shims in between each plate, and then soldered the stack to a brass shaft. When I removed the shims, the plates were more or less evenly spaced. (Figure 14-8)

Yet, despite several hours of tinkering, I was never able to get the plates aligned well enough to make my rotor assembly useable. In principle, I still think that building a workable rotor this way is possible. What's required is some sort of fixture to hold the plates in perfect alignment while they are being soldered to the shaft.

Figure 14-8
A homemade brass rotor

Chapter 15
The "Roofing Metal" Condenser

We left the last chapter with a description of the rotary condenser, and my first failed attempt to build one from scratch. Happily, my second attempt was more successful, but only after reconsidering the essential characteristics of the rotary condenser itself. In this chapter, I would like to share construction details for a practical, home-built rotary condenser. (Figure 15-1)

Previously, I had tried to fabricate a rotor with rigid brass plates. The intent was to mesh the rotor with equally spaced stator plates, and use air as the dielectric. Try as I might, I could not assemble the plates with enough precision to prevent one or more rotor plates from touching the stator plates.

In analyzing my failure, it occurred to me that rigid plates were not an essential characteristic of the design, but an optional one. Rigid plates are only needed If air is used as the dielectric. Rigidity provides mechanical stability, which ensures a uniform gap between the plates.

If I changed the dielectric to a plastic sheet or film, precise alignment of plates would no longer be an issue. No matter how poorly aligned the plates might be, they could never touch.

In this case, rigid plates were actually undesirable. I realized that thin, compliant rotor and stator plates would be much better, because they could flex to absorb the effects of poor alignment or lack of mechanical precision. The result is the design that follows.

Let's begin with the plates. Stator plates were fabricated from aluminum roofing metal disks, 4 inches in diameter. A one half inch hole was cut into the center the stator plate. Next, I cut a semi-circular window with a radius of about 1 1/8". Finally, I drilled four 1/8" holes,

Figure 15-1
The "roofing metal" condenser

equally spaced, around the edge of the disk.

Next, came the rotor plates. They are basically semi-circles, with a radius of 1 ½ ". Notice the "hub" at the center of the plate. The hub is about ½" in diameter, and was drilled with a 1/8" hole.

If you plan on making many plates, it pays to first make some cardboard templates. Once you have a template, it's easy to transfer lines and hole locations to the aluminum. This is true for both rotors and stators. (Figure 15-2, and Figure 15-3)

Dielectric disks were fashioned from sheet plastic, purchased at a local office supply store. The plastic is sold for use on overhead projectors. The dielectric disks, like the stator plates, were cut to 4" diameter. A common office hole-punch was used to punch four equally-spaced holes around the perimeter of the disk. A 1/2" hole must be cut in the exact center. I used a piece of 1/2" pipe, warmed with a torch, to melt perfect holes into the center.

A simple condenser requires three stator plates, two rotor plates, and four dielectric disks. It's easy to build condensers with more plates, but let's stick to this case first. Once you understand how it is

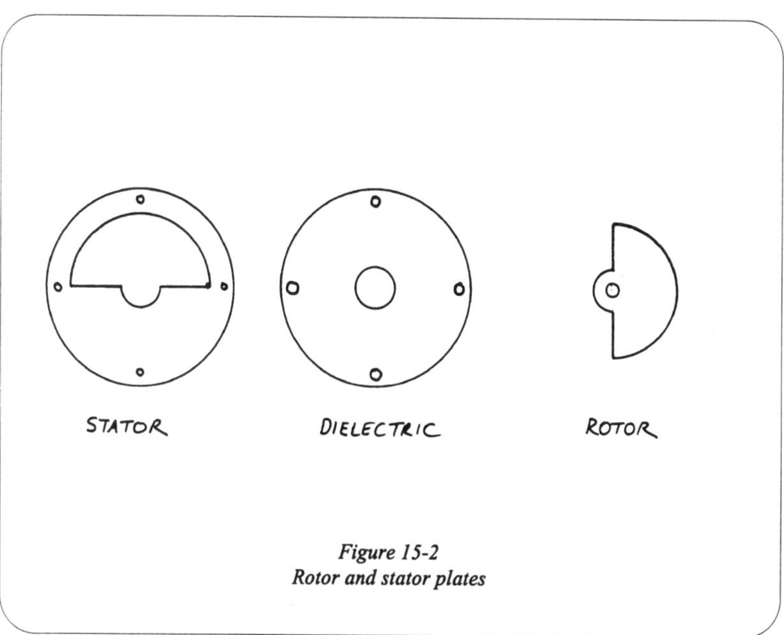

Figure 15-2
Rotor and stator plates

assembled, you can always add more plates later.

Assembling the condenser requires everything to come together at once. I began with a stator plate. I inserted a 1 1/2" length of 6-32 threaded rod into each hole along the stator's perimeter, and backed them with 6-32 nuts.

Next, I laid down a dielectric disk. The threaded rods were guided through the four holes in the plastic disk. A lock washer was then slipped onto each protruding rod. The dielectric disk was adjusted as necessary, so that the lock washers on each rod was centered within the punch holes in the plastic.

Next came the rotor plate. The rotor plate was laid on top of the dielectric disk. The rotor shaft, a 2 /12" length of 6-32 threaded rod, was inserted through the rotor.

The second dielectric disk was added to the sandwich. Again, it was adjusted to make sure that it's perimeter holes lined up with the lock washers on the threaded rods. I added the stator number two, followed by another dielectric disk. Lock washers were placed on each perimeter rod.

A lock washer was also threaded onto the rotor shaft, followed by rotor plate number two.

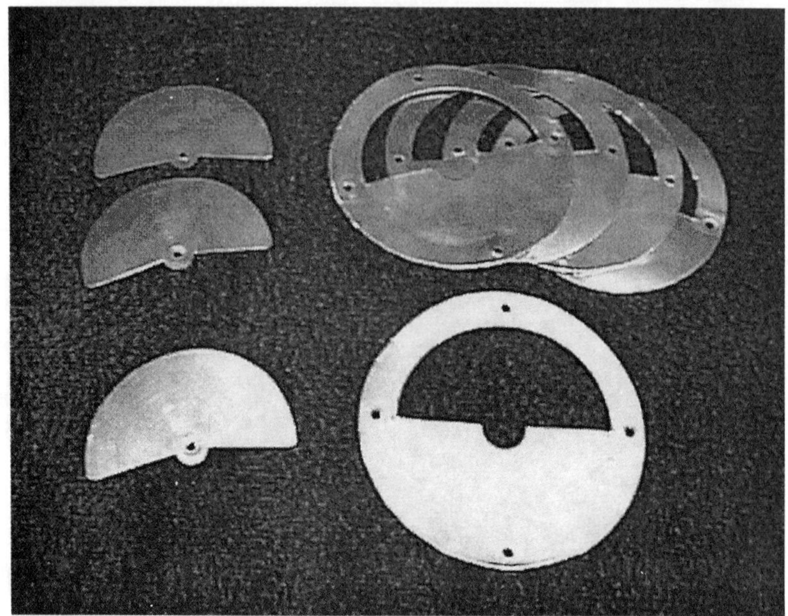

Figure 15-3
Rotor and stator plates

The last dielectric plate was added to the stack followed by the final stator plate. The perimeter rods were capped with lock washers and nuts, and the entire sandwich was tightened. The rotor shaft was fitted with lock washers and nuts on each end and tightened to fix the rotor plates to the shaft.

The lock washers on the perimeter rods serve two functions. First, they space the stator plates enough to allow two dielectric sheets and one rotor plate to fit between them comfortably. When the rods are tightened, the lock washers bite into the stator plates and connect them all, electrically.

The lock washers on the rotor shaft serve a similar purpose. They space the rotor plates uniformly, lock them to the shaft so that they will rotate when the shaft is turned, and link them electrically.

Figure 15-4 shows an exploded view of the relationship between the stators, dielectric sheets, and rotors. Figure 15-5 shows some details related to the rotor and shaft assembly. Remember, however, that the entire assembly is built up at once.

During assembly, the rotor plates must be lined up. You may

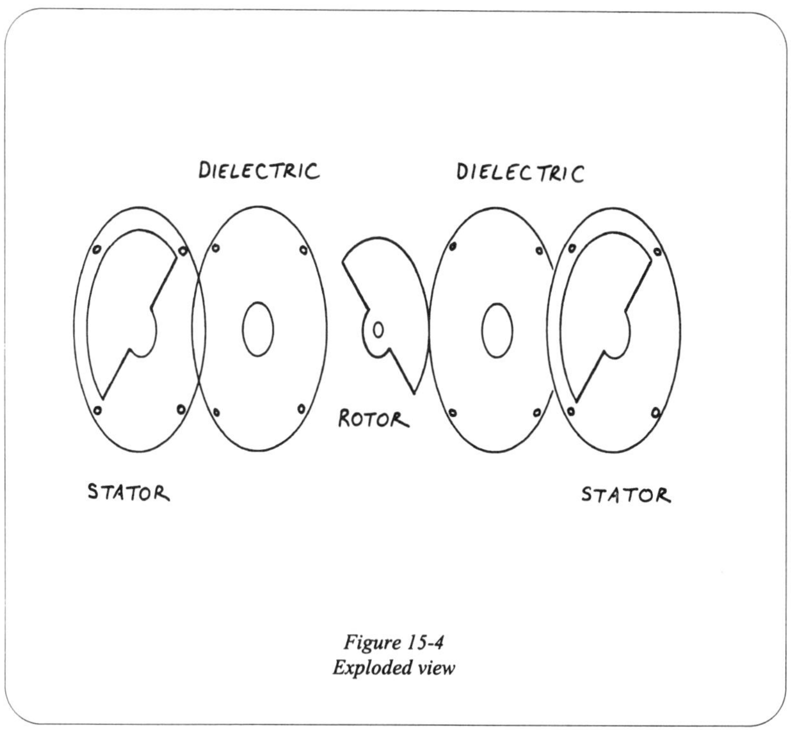

Figure 15-4
Exploded view

wonder how I accomplished this considering that, once they are buried between sheets of plastic, you can't touch them. The answer is simple. I pushed a common pin through the stator window, and through all the layers of dielectric. I tightened the nuts on the rotor shaft with my fingers, and then rotated the shaft clockwise, until the each rotor plate had struck the pin. With the plates all aligned, I tightened the rotor nuts real hard with two wrenches. Finally, I yanked the pin out. All that was left was a very tiny hole that doesn't seem to make any difference. (Figure 15-6)

Adding more rotors and stators is simple, if you remember a few rules. Always begin and end the stack with a stator plate. Always make sure that there is a dielectric sheet on both sides of a rotor plate. Finally, use lock washers between adjacent plates, whether they are rotor or stator plates.

The next step was to prepare a base for the condenser. A suitable piece of wood was drilled with four holes, aligned with the perimeter rods on the stator plates. Rods were slid into the holes, which

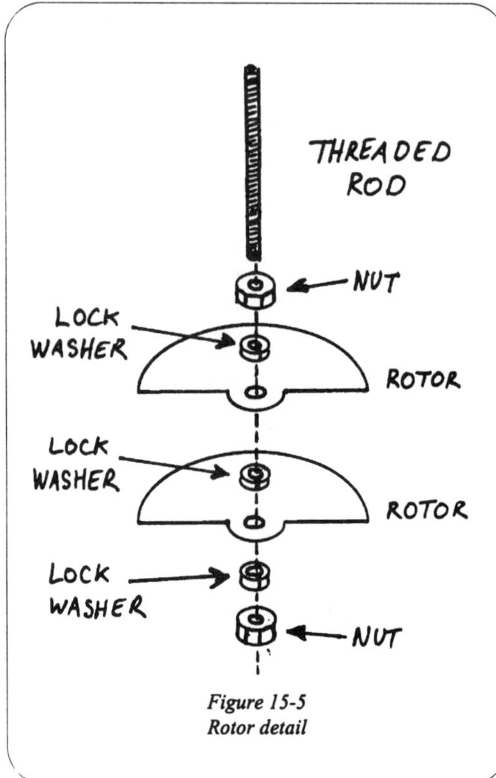

Figure 15-5
Rotor detail

were counterbored from the rear to accept 6-32 nuts. When tightened the condenser stack was anchored securely to the base.

A fifth hole, centered between the other four, was drilled to accept the back end of the rotor shaft. In my case, I found a tiny brass bushing in my junk box. I drilled the center hole in the base to fit the bushing, and pressed it into the hole. The rotor shaft was inserted into the bushing, which provided a smooth, low-friction bearing. (Figure 15-7)

After bolting the condenser stack securely to the base, I selected two opposing perimeter rods, and trimmed them as far as possible. The other two I left long.

I cut a strip of pine, 4 1/4 " long, 3/4" wide, and 3/8" thick. This was intended to serve as the upper bearing block. Three holes were drilled into the block, one near each end, and one in the middle. The end holes were spaced to align with the uncut perimeter rods, while the center hole was drilled to fit the rotor shaft.

The block and the two perimeter rods that support it comprised a "gallows" type arrangement, with which you should now be familiar. The upper bearing block can be seen in figure 15-7, and figure 15-8.

Prior to assembly, a short length of brass tubing was pressed into the center hole of the bearing block. The tubing was slightly longer than the block was thick, which allowed it to project a bit above the wood. This brass tube provided a smooth bearing surface, and a moveable electrical connection to the rotor.

The wooden base was drilled for two 6-32 bolts with nuts and wing nuts, set up as binding posts. One binding post was connected to the stator stack, by attaching a wire to one of the perimeter rods. The other binding post was linked with wire to the brass bearing tube in the upper bearing block. The wire was soldered to the body of the tube.

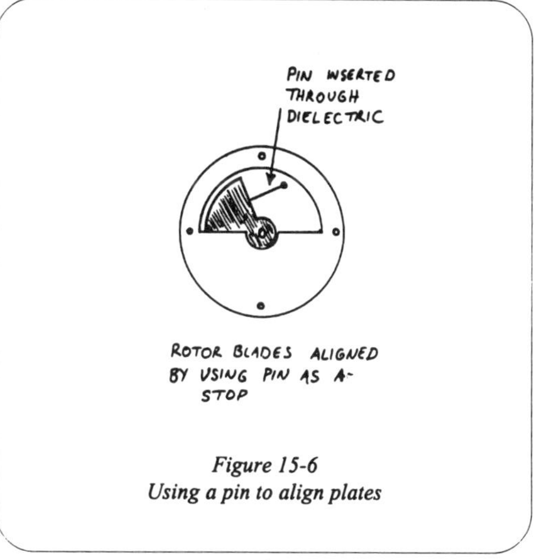

Figure 15-6
Using a pin to align plates

As is always the case, once all condenser components were fitted and functional, the condenser was removed from the base, which was then sanded, stained, and polyurethaned. For an authentic look, the wires linking the binding posts to the condenser were "cotton-covered" using some old, tubular shoe laces. Finally green felt was cemented to the bottom of the base.

End Results and Going Farther

After it was completed, my condenser received a few additional enhancements. For starters, I found the 6-32 rotor shaft somewhat difficult to manipulate. I wanted some way to "thicken" the shaft to make it easier to grasp. One approach was to slide a length of copper or brass tubing onto the shaft and solder it into place. I used an aluminum standoff.

Standoffs are small metal cylinders, which generally run from 1/4" to 2" in length. They are often 1/4" in diameter, with a round cross section, though some are hexagonal. The center of the standoff is generally bored through, and may be internally threaded. Standoffs are intended for mounting circuit boards to the chassis or housing of a piece of equipment. They can be found in electronic junk, and from electronic

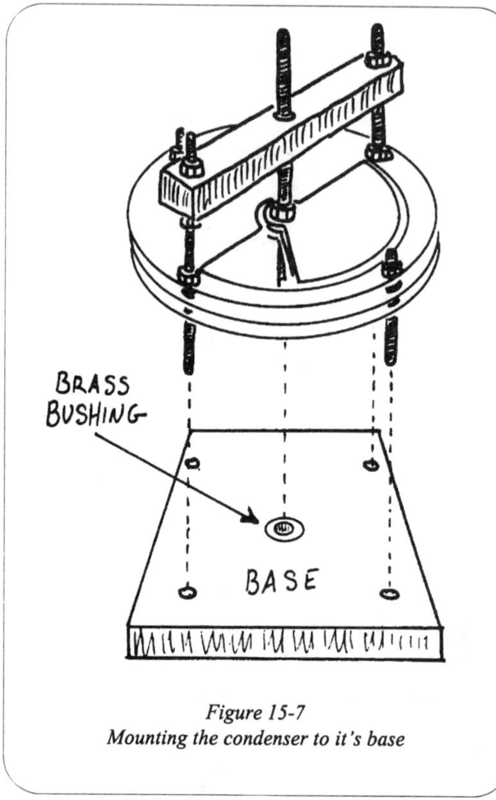

Figure 15-7
Mounting the condenser to it's base

supply houses, both new and as surplus.

The standoff I found was 1 ½" in length, 1/4" round in cross section, and threaded internally for a 6-32 bolt. It screwed right on to my rotor shaft. Once the rotor shaft was modified, the upper bearing had to be enlarged. I found another piece of brass tubing that fit smoothly over the standoff, and it provided an excellent bearing. I got fancy with the upper bearing, and instead of simply forcing it into the block, I soldered a flange to the tube. The tube was slid into the block, and the flange was bolted down with two 6-32 screws and nuts.

Expanding the rotor shaft to 1/4" allowed me to attach a standard radio knob. I had a junk box full of them to select from. This made the condenser even easier to adjust. Of course, a knob with a pointer implies some sort of scale. This was added to the condenser using a cheap plastic protractor. The protractor was drilled to accommodate the rotor shaft and bolted to the bearing block using the same bolts that anchored the bearing flange to the block. (Figure 15-9)

Using an array of four stator plates and three rotor plates, with the approximate dimensions given, produces up to 600 picofarads when fully meshed. The knob and the protractor scale allows the condenser to be adjusted, and then returned to a previous setting with considerable repeatability.

I connected a capacitance meter to the condenser, measured the capacitance, and noted where the knob was set. I gave the knob a random whirl. Then, I returned it to it's previous position and measured the capacitance again. In fact, I repeated the process several times, and discovered that the condenser is repeatable to within a two percent of full scale!

Here's an observation: In circuits which require one terminal of the condenser to be grounded, I prefer to ground the stator terminal. I've noticed that the outer stator plates provide some amount of shielding. This helps reduce the effects of a problem called "body capacitance". Body capacitance can make a tuned circuit go loopy when your hand approaches it.

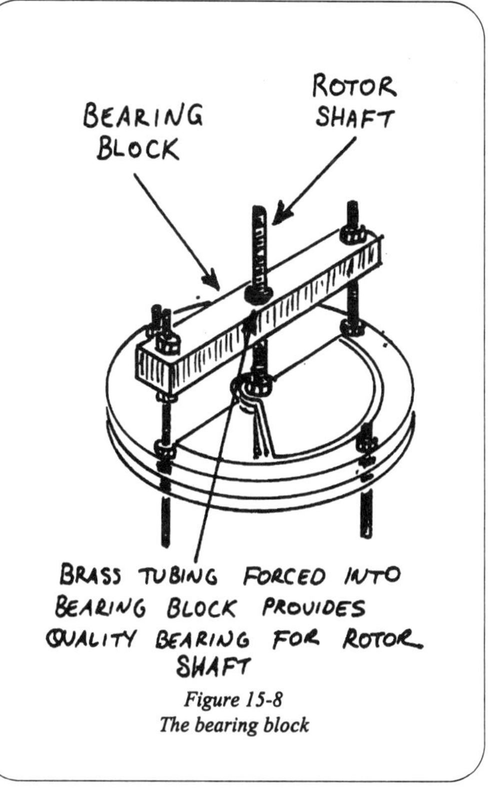

Figure 15-8
The bearing block

Where do you go from here? Despite the fact that this design makes for a decent piece of homebrew equipment, there is always room for experimentation and possible improvements.

What other plate materials might be available? It has occurred to me that aluminum from a common beer or soda can might make an excellent plate, given it's thinness and it's relative flexibility.

What other dielectric materials would be suitable? Can you think of a better means to assemble and lock rotor plates on to a shaft? I'm betting that there is. Another thought: I've wondered if there would

Figure 15-9
The bearing block and the dial

be any mechanical or stability advantages to using more than four perimeter rods.

Finally, consider the shape of the rotor blades themselves. As semicircles, they produce a capacitance value that is linearly proportional to the shaft rotation. Is that necessarily a good thing? What other rotor shapes might have offer advantages? "Stay tuned." You may be surprised...

Chapter 16
Essential Characteristics Of The Coil

In the preceding sections, I introduced condensers as a key component of electrical tuning systems. Earlier than that, I mentioned that resonant circuits (electronic "tuning forks") also depend upon a second vital component, the coil. This chapter is dedicated to the introduction of the coil, and it's essential characteristics.

Let's begin by considering a large, heavy iron flywheel. If we attach a crank to it and whirl it around, we can cause it to rotate. Note that the wheel can't spin itself, rather, it has to be set into motion by an external force. In this case, energy produced by your muscles is transferred through the crank to the flywheel. The energy from your effort hasn't disappeared; it's now stored in the motion of the wheel.

Once energy has been stored in the wheel, it can be withdrawn at any time by engaging some gears or a belt to link the flywheel to whatever machinery we want to power. If you continue to draw energy from the wheel, it's energy will eventually be drained, and the wheel will come to a stop.

Toy manufacturers use to sell toy cars powered with flywheels. You'd rev up the flywheel by raking the car across the floor two or three times. Your effort was stored as mechanical energy in the flywheel. When the toy was released, the flywheel transferred the stored energy to It's wheels, and the car would drive for a considerable distance under it's own power.

If we take an iron or steel nail, wind many turns of insulated wire around it, and then connect the wire to a battery, current will flow through the wire and produce a magnetic field. The field will concentrate, appearing in the iron nail. Like the flywheel, which requires energy or effort to set it into motion, it takes energy to build up that magnetic field. The energy taken from the battery doesn't vanish, rather, it's represented by the field that is created.

If the battery is suddenly disconnected, the magnetic field cannot sustain itself. It will begin to collapse, but in doing so, it will transfer some of it's stored energy back into the wire! Just as the flywheel, which can accept energy and later give it back, the coil can accept, store, and then release electrical energy. A coil, like a flywheel, is an energy storage device.

The actual process by which a coil stores energy is called *inductance*. The greater a coil's inductance, the greater it's capability to store energy. This is similar to the flywheel in the sense that a heavy flywheel can store more energy than a small, light one.

Inductance actually exists in all conductors, even short, straight pieces of wire. Technically speaking, all conductors are *inductors*, although that term is usually reserved for devices intentionally designed to produce specific inductance values.

Coils represent a specific family of inductors. Practically speaking, all coils are inductors, but not all inductors are coils. In crystal radio sets, inductors almost always appear in the form of some type of coil.

Coils and inductors have another interesting property. Like condensers, their ability to pass electric currents is frequency sensitive. However, their reaction to specific frequencies is entirely opposite that of the condenser.

Condensers, if you recall, favor quickly alternating currents. If the frequency of the current passing through the condenser is reduced, the condenser's impedance increases. For steady, direct currents, like those from a battery, the condenser's impedance goes to infinity, and it won't pass the current at all.

The coils, on the other hand, will easily pass direct currents. They will also pass slowly changing or oscillating currents. However, as the frequency of the current increases, the coil's impedance also increases. The greater the frequency, the greater the coil's impedance. Coils favor slowly alternating currents. This prejudicial behavior against higher frequencies is called *inductive reactance*.

"High frequency" is a relative term. The frequency at which a signal can no longer pass through the coil depends on it's inductance. A coil with a small inductance value will pass higher frequencies than a

$$L = \mu \frac{N^2 A}{l}$$

Figure 16-1
The formula for single-layer coil inductance

coil with a large inductance. In other words, the coil's reactance is dependent upon it's inductance. It is the complementary nature of capacitance and inductance that makes condensers and coils function as tuners when they are combined.

Figure 16-1 shows the formula for inductance in specific type of coil. The coil is tubular, and is neatly wound with a single layer of insulated wire.

"L" is the mathematical symbol for inductance. Inductance is measured in units called *henries*. Large value coils exist for various purposes, but for crystal radio tuners and similar circuits, inductor values tend to be quite small. Coils for those applications are typically measured in millihenries (thousandths of a henry) or microhenries (millionths of a henry).

The "N" term in the equation is the number of turns of wire on the coil. I said earlier that all conductors, even straight pieces of wire, have some amount of inductance. That inductance is related to the magnetic field that circles the wire when it is energized. Looping wire in to a coil causes the field around any particular segment of wire to combine with the magnetism from adjacent windings.

The focusing effect of multiple windings is very powerful. Doubling the number of turns on a coil can actually increase the coil's inductance by a factor of four, provided none of the other variables are changed.

The "A" term refers to the area of the cross-section of the core, or the area enclosed by one turn of wire. Inductance is directly proportionate to the core area. All things being equal, if you double your core area, you will double the inductance of the coil..

The "l" term refers to the length of the coil. Coil length actually works against inductance. When you spread out a coil's windings over a greater distance, you degrade the ability of each turn to influence adjacent turns.

The strange-looking character in the formula is the Greek letter mu, which represents the permeability of the core material. Certain materials are sympathetic to magnetic fields, and when exposed to them, exhibit a focusing or enhancing effect. A coil of a certain length, number of turns, and core area, wound on a steel core may have an inductance many thousands of times greater than an identical coil wound on a wooden core.

The choice of conductor and core material can affect aspects of the coil other than the inductance. Recall that in the flywheel example, muscle energy was used to set the wheel in motion. Later, when a load was connected to the flywheel, we got all of the energy back.

In reality, flywheels rotate on bearings, and no bearing is perfect. Friction is always present. Furthermore, air in contact with the wheel causes drag. Energy is bled away from the wheel to overcome those losses, which cause the wheel to gradually slow down. In fact, if we wait long enough, the wheel's stored energy will eventually be consumed. The wheel will grind to a halt, and all of the energy you expended setting the flywheel in motion will have been lost.

Coils are real-world devices, and they suffer from similar losses. If we store energy in the magnetic field of a coil, and then try to recover it, we never get it all back. A certain percentage is always lost. The trick is to select your core and conductor materials so that these losses are minimized. For radio work, copper wire and air cores are a good place to start. Trial and error can take you from there.

The inductance of two or more coils can be combined by simply connecting all of the coils in series. The resulting inductance will equal the sum of the inductance values for each coil in the string. If two coils are connected in parallel, the resulting inductance will be something less than the value of the coil with the least amount of inductance.

Be careful when you place two tuning coils near each other. The inductance of each coil is the result of the action of a magnetic field. Each coil has it's own field, which radiates from it. If the field from one coil joins the field of the second in a sympathetic way, the inductance of the combination may be much higher than expected. Similarly, if the fields from the two coils clash, the total inductance may be much lower than predicted. If you have to mount two coils next to each other, and want to minimize their interaction, mount them so that the axis of one coil lies at right angles to the other.

We've come a long way in this discussion, so let's pause to highlight the essential characteristics of the coil.

First, all conductors exhibit a property called inductance. A device specifically designed to produce inductance is referred to as an inductor, and a coil is a specific type of inductor. Inductors and coils have the capability to store and release energy. They also have the ability to discriminate between electrical currents with different frequencies.

In the case of a simple single-layer coil, it's inductance depends on the number of turns of wire, the length of the coil, the core material on which the coil is wound, and the area of the cross-section of the core. The difference between a good coil and a better coil has to do with it's tendency to lose energy. Losses depend upon what materials comprise the wire and the core.

Finally, if two or more coils are connected in series, they create a "super" coil whose inductance value equals the sum total of all of the contributors. Coils radiate magnetic fields, and two coils, placed near each other, can interact.

Page 118

Chapter 17
Thoughts On Practical Coils

Coil building can be a lot of fun. It's not overly complicated, but having a few tips to start off with can make the process much easier. Once you've wound a few, you can develop your own style and technique, and a stable of your favorite designs. Let's begin with a single layer, tubular coil like one in figure 17-1.

Wire is your first consideration. I think that it's safe to say that you could wind suitable coils out of just about any kind of wire, though some types work better than others. For instance, some wire has a solid center, while other types are composed of strands. Stranded wire is designed to be flexible. I prefer solid wire, because it's easier to wind, and holds it's shape better.

Wire comes with different types of insulation. Bare wire is available, but not suitable for winding most coils. If one bare winding touches the next they "short", and the inductance effect is muted. Unless your coil design can prevent one turn of wire from touching it's neighbors, stick to using insulated wire.

Some wire is insulated by coating it with enamel. It almost looks like bare wire, but if you scrape it with the edge of a knife, the coating flakes off exposing shiny copper underneath. Enameled wire is used extensively in commercially manufactured coils of all types. Enamel is a cheap and fairly durable coating that doesn't significantly increase the diameter of the conductor. This is important, because it means that a tightly-wound coil of enameled wire won't be much larger than an equivalent coil wound from bare wire.

Enameled wire is easy to get. Most electronic supply houses carry it. I prefer to salvage the wire I need from electrical and electronic

Figure 17-1
Various handmade coils and spools of magnet wire

junk. Any electrical appliance that contains coils is a potential goldmine. Old radios, televisions, tape and record players come to mind. Motors, old doorbells, and wall transformers are also good candidates. An old pair of electric hair-clippers produced a nice spool of excellent wire. All of these devices depend upon the production of magnetic fields for their operation. For this reason, enameled wire is commonly called *magnet wire*.

When you tear down electrical junk in search of magnet wire, expect to find it in two forms. Sometimes, you'll find it wound on some sort of plastic spool. This is ideal, because all you need to do is figure out how to get the spool out. The outside of the spool is sometimes wrapped with a plastic or cloth tape, so you'll need to peel that off before unwinding it. The adhesive from the tape will probably have contaminated the first layer of wire, so I usually pull that much off and discard it.

Other times, the coil is an irregular shape, and there is no easily removable spool. Yoke coils, wound to fit the neck of picture tubes fall into this category. In that case, you have to carefully unwind the wire, and transfer it to a spool of your choice.

Enameled wire comes in different colors, depending on the enamel used. I've wound coils in red, brown, and green. I even found one instance of clear enamel! The wire looks like it's bare copper, but it's not!

The most common wire you are likely to come across will have plastic insulation. Plastic coatings tend to be thicker than enamel as an insulating material, which means you can't wind as many turns of wire in a given space.

Moreover, some coils are built using techniques that work better with enameled wire than plastic. I have wound coils with plastic-coated wire, but I prefer to use magnet wire. Try them both and be your own judge.

Wire gauge is also an issue. I've seen magnet wire that was so thin that a human hair looked like a log next to it! Conversely, I've seen samples so thick that you needed two hands to bend it.

Choosing wire gauge for a home-built crystal set is largely a matter of common sense. Extremely fine magnet wire is difficult to manage. It has a

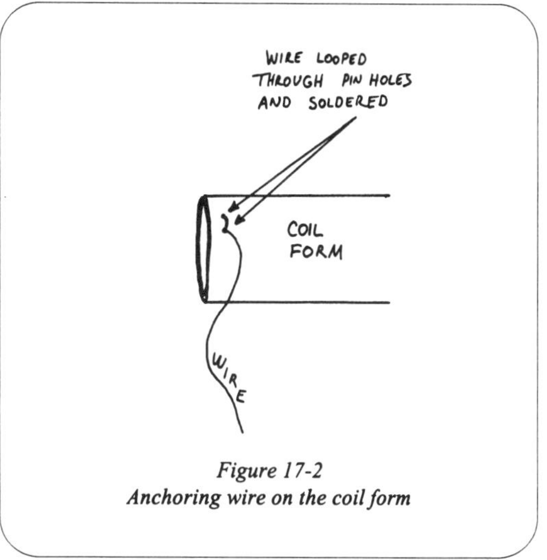

Figure 17-2
Anchoring wire on the coil form

Figure 17-3
Winding by hand

tendency to tangle and break when you work with it. It's also very difficult to wind onto a form in a neat, single layer. It's probably ok for so-called scramble-wound coils, but forget it for single-layer coils.

If you purchase magnet wire, the seller will want to know what gauge you want. I would suggest that you buy several different gauges and gain some experience will them. Be warned that wire gauge numbers are counter-intuitive. The larger the gauge number, the *thinner* the wire. If you want a guideline, start with something in the low 20's.

Your next concern is the choice of core. Most tuning circuits for crystal-type radios require coils with fairly small inductance values. Generally, this boils down to using air as the core material. Since it's difficult to wind a coil on air, thin tubular forms are used. If the proper form material is selected, it's presence doesn't affect the inductance to any significant degree. For all practical purposes, the coil is still an air-core coil.

Paper and cardboard tubes make excellent coil forms. You can fabricate your own tube, but the average garbage can is an outstanding source of forms. An obvious coil form appears as an empty toilet paper or paper towel tube.

The tubes from the center of gift wrapping paper, or the tube from the center of a roll of fax paper are also good choices.

I especially like cardboard mailing tubes, because they have very strong walls, and are smooth and well finished on their exterior. Then again, I wish I had a dollar for every tuning coil wound on an empty oatmeal box throughout radio history.

Common PVC pipe is often used by experimenters. PVC is inexpensive, stable, light, and easy to cut and work with. I've even wound coils on glass jars and bottles.

The winding process for tubular single-layer coils is largely a matter of personal technique. Generally, I begin by punching or drilling two tiny holes in the form. The holes are about 1/4" apart, and about 3/8" or 1/2" from the edge of the form.

I take a strand of magnet wire, and strip the insulation about 2". This can be done by carefully scraping it with a knife, or sanding it with some emery or fine sandpaper. The stripped end is threaded through the two holes, back and forth, two or three times. This forms a nice anchor for the start of my windings. (Figure 17-2)

I like to touch the anchor loop with a soldering iron and tin the wire with a nice, shiny dot of solder. It's a good idea to drill two more anchor holes at the other end of the tube. These will be used to anchor the other end of the wire when the winding operation is complete.

A spool of magnet wire is placed on the shaft of a screwdriver, which has been clamped in a vise, to provide an axle so that the wire can unwind without difficulty. The coil form is held in the right hand (I'm right-handed). The index finger of the left hand is placed underneath the form. The thumb presses the magnet wire against the top of the form. The remaining fingers are curled around the magnet wire.

Using the right hand, the form is slowly rotated clockwise. The thumb is used to guide and align the wire as it is laid down on the form. The index finger provides pressure, and the remaining fingers help keep the wire under uniform tension. (Figure 17-3)

When sufficient windings have been added to the coil, the end of the wire is stripped, threaded through the anchor holes, and soldered. Figure 17-4 shows just such a coil, wound on a mailing tube 2" in diameter and 4" in length.

Winding can be tedious. One way to speed the process is to install a plug or cap in each end of the tube. The plugs might be rubber stoppers, corks, pipe caps, or even the plastic lids that come with mailing tubes. A 1/4" hole is drilled in the center of each plug, and a

Figure 17-5
An example of a tapped coil

piece of 1/4" threaded rod is slid through the core. A couple of nuts on each end secure the rod.

Next, the rod is chucked into an electric drill. The drill must be the variable-speed type, with as fine a control as possible.

I have two drills, both of which are variable speed units. One provides precise control all the way down to zero motion. It's an exceptional tool with which to wind coils. The response of the other is somewhat coarse. It's impossible to control it with enough precision to keep from breaking or knotting the wire. Some drills work well for winding coils, some do not.

Please exercise extreme caution when winding! Magnet wire is thin, and if drawn quickly over the skin, it will cut into your flesh as well as any knife.

Whether the coil is wound by hand, or with the aid of a drill, I like to seal all finished coils with a coat of polyurethane. The coating provides strength and stability, as well as protection. In coils where the wire ends are not anchored, a coating of polyurethane will prevent the coil from unwinding.

If you insist on building a true air-core coil, I would suggest the following technique:

First, locate a straight-walled jar or bottle with a diameter equal to the diameter of your intended coil. Wrap the waist of the bottle with a sheet of waxed paper, and fasten it with a small piece of tape. Carefully wind your coil over the waxed paper, placing as many turns on the bottle as you wish. The coil should be neat, so that each turn of wire touches the turn before it. The free ends of the coil can be anchored with bits of tape.

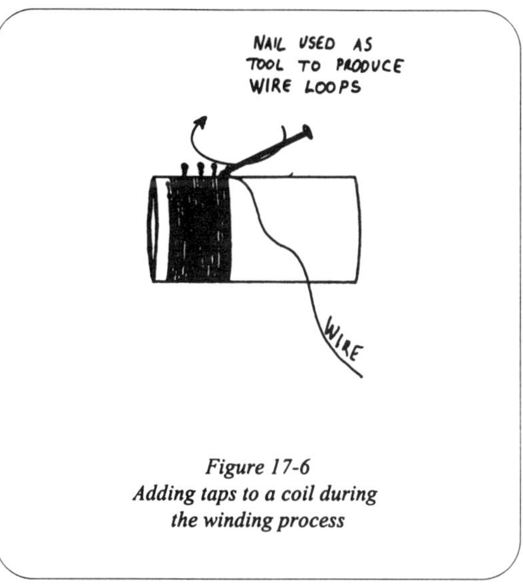

Figure 17-6
Adding taps to a coil during
the winding process

When you're finished, give the coil a liberal coat of shellac or polyurethane. Once the coating has dried and hardened completely, add a second coat. The purpose of the coating is to bond the wire into a solid mass.

When the second coat is dry, it should be possible to gently remove the coil from the bottle by sliding it off. The waxed paper should peel away, leaving a solid, if fragile, wire cylinder. I would dip

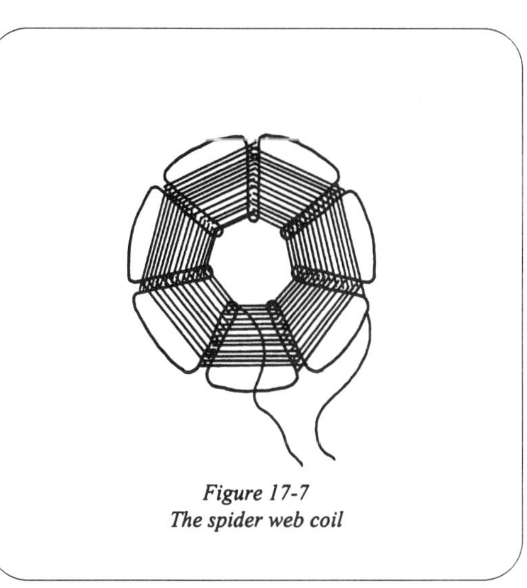

Figure 17-7
The spider web coil

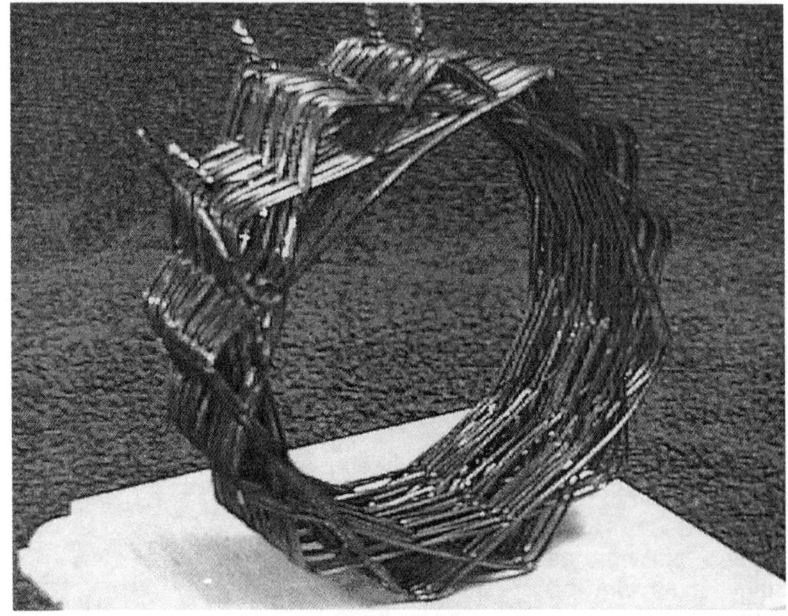

Figure 17-8
A basket weave coil

the entire coil in shellac or polyurethane once more for good measure, and hang it up to dry. A coil of this type should cemented to a suitable base with wing nut terminals for easy connections.

Each of the previous coils provides a specific value of inductance. If you need more or less inductance, you have to wind a different coil. One way around this inconvenience is to wind your coils with taps. Figure 17-5 shows an example of a tapped coil.

Taps are electrical connections that are placed at regular intervals throughout the windings, They allow you to electrically add or subtract windings to produce an inductance value that will more closely match your needs.

Winding a tapped coil is much like winding a plain one. After the anchor has been set and ten or fifteen turns of wire have been laid onto the form, I pause. Using a small nail, I introduce a tap by twisting a little loop in the wire. The little loop stands up, perpendicular to the side of the coil. (Figure 17-6)

Next, I add another ten or fifteen turns to the coil, and produce another tap. The winding and tapping process is repeated at regular intervals across the length of the coil. When the winding process is complete, the wire is terminated in the usual fashion. The entire coil is coated with polyurethane. When the coat is dry, the taps are stripped of insulation and soldered.

All the coils described above have a common architecture: multiple turns, side by side, in a single layer. It is possible to build coils by rearranging these relationships. Imagine multiple layers, one on top of the next, built on a *single turn* of wire. An example of this is the so-called spider web coil.

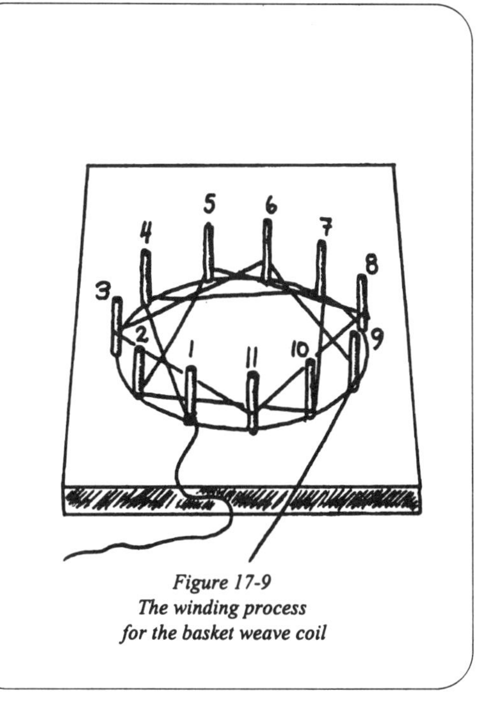

Figure 17-9
The winding process
for the basket weave coil

Spider coils begin with flat, rigid forms. I've seen forms made from heavy cardboard, thin wood, masonite, or plastic. The forms are frequently circular in shape. A series of slots (an odd number), laid out like the spokes of a wheel, are cut into the form. The slots run from the near-center of the form all the way to the edge.

To wind a spider coil, a strand of wire is laid into the bottom of one of the slots, running from the front of the form to the back. The wire is stretched to the next slot, and passes from the back to the front. At the third slot, it's threaded from front to back again. This weaving process continues around the form until enough wire has been accumulated to fill the slots. The resulting coil is more or less flat, with a geometric pattern that suggests the work of a spider. (Figure 17-7)

Like the tubular coils, spider coils can be tapped. Unfortunately, the inductance equation given for the tubular coil doesn't accurately describe the behavior of spider coils.

Figure 17-8 shows an interesting coil type. I've been told by several people that it looks more like a Christmas ornament than a radio

Figure 17-10
A loop antenna

component. Some people call it a basket weave coil. Why would anyone go to the trouble of making coils this way?

When I introduced condensers, I made a general statement to the effect that anytime two conductors lie near each other separated by an insulator, you've created a condenser. Believe it or not, each pair of neighboring turns of wire on a coil can have capacitance in addition to their inductance! This capacitance is an unintended side-effect of coil-winding, and in some cases, it causes problems. The basket weave design is intended to produce usable inductance values with low capacitance.

To wind a coil like this, you need a scrap piece of pine board, clamped face up in a bench vise. Using a compass, I lay a circle on the wood, and then divide the circumference of the circle into eleven equal points.

Next, I secure eleven box nails. One nail is pounded into the board at each point along the circle. The nail isn't driven into the wood more than a 1/4". If the nails have heads, I use nippers to cut them off. The resulting ring of nails provides a form on which to "weave" the coil.

Weaving is easy. I choose one nail as a start, and wind some magnet wire around it a few times to provide an anchor. Assuming the nails are numbered clockwise, and that my anchor starts at nail #1, I stretch the wire through the center of the circle to nail #4. The wire is stretched around the outside of nail #4, and back into the center. Next, it's bent around the outside of nail #7, and back to the inside of the circle. The wire is threaded around nail #10, back into the circle, and then around #2. (Figure 17-9)

The weaving pattern is simple: the wire passes around one nail, then inside of two, around one nail, then inside of two. The process is repeated over and over again. You can throw down as may layers of wire as you want, though you are limited by the height of the guide nails.

When the winding is complete, the coil has to be stabilized and stiffened. To do this, I paint the windings with polyurethane, taking care not to get the nails wet. Once the coating is dry, the trick is to carefully life the coil basket off of the nails. If you succeed in doing this, I'd recommend immersing the entire coil in the urethane to further coat and strengthen it.

In a normal coil, each turn of wire is in close proximity to it's neighbors for it's entire length. In the basket weave design, turns only contact adjacent turns at the points where the woven wire intersects itself. For the rest of it's length, it is separated from the adjacent windings by air. This reduces the effective area of the adjacent windings, which reduces the unwanted capacitance.

Not all coils are small devices. Loops are large coils, typically wound around a wooden or plastic frame of some sort. The largest I've ever read about was wound around the frame of an old door, though most of the designs I've seen keep them in the range of one to three feet in diameter.

The simplest frame design is nothing more than two wooden strips, arranged and fastened as a cross. The end of each leg of the cross is notched. Wire is laid into the notches, and is wound around the outside of the frame. (Figure 17-10)

Loops are often used as antennas. An interesting feature of the loop is that it tends to be directional. By aiming the loop, it's possible to encourage the reception of some signals, and discourage the reception of others. It can and has been used to determine the unknown location of a transmitter.

Air, of course, is not the only useable core material. Sometimes, higher inductance values are useful, and an air-core coil is simply too large. Or, a designer may wish to use less turns of wire, yet maintain a given inductance. The solution is to introduce a core material with a higher mu.

Iron and steel have larger mu's than air, but they make poor cores for radio work. I think they are simply too lossy. There are, however, many ceramic-like materials that offer high mu's with low losses.

These materials are known as ferromagnetic materials, or ferrites. They are highly magnetic, very hard, and actually rather brittle. They'll shatter if dropped on a hard floor.

Cheap transistor a.m. radios feature a coil called a *loopstick*. The loopstick is nothing more than a short bar of ferrite material with one or more coils wound around it. In other applications, the ferrite material is molded into a donut-shape, with coils wound around the body. Donut-shaped cores are more properly known as toroids.

Lots of cheap radios still use loopsticks. They are there for salvage if you are so inclined. Toroids appear in a lot of electronic junk these days. Look for them in old computer power supplies, or in junked televisions and VCR's.

As you can see, there is an almost endless variety in coil designs. Given the ample availability of materials, there's no reason not to try them all.

Chapter 18
Practical Variable Inductors and Coils

The coils described in the previous chapter are all considered "fixed" coils, in that their inductance values can't adjusted. Sometimes a circuit demands different inductances for different circumstances. This might require maintaining a large assortment of coils, swapping them in and out as necessary. A variable inductance is a better idea, and a tapped coil would seem to be a good candidate for the job.

Yet, while tapped coils are adjustable, they are not truly variable. The increase in inductance from one tap to the next is not smooth and gradual. It is a fixed, defined step, which depends on the spacing of the taps and the number of turns of wire between each pair. If the inductance of one tap is too low, and the next is a little too high, you're out of luck. There is no in-between setting.

If you add more taps, and reduce the number of turns between adjacent taps, you can reduce the incremental change from tap to tap and achieve finer resolution in your adjustments. The problem is that there are practical limits to the number of taps you can throw into the winding process. Fabrication becomes complicated as well. How, then, can a coil be engineered to be variable?

I've already introduced the inductance formula for a single-layer coil. If you recall, that equation said that the coil's inductance was affected by the number of turns of wire in coil, the coil's length, the area of it's core, and the material in the core. If we want to build a variable inductor, all we need to do is devise a scheme to vary one of these important parameters. The essential characteristic of a variable inductor

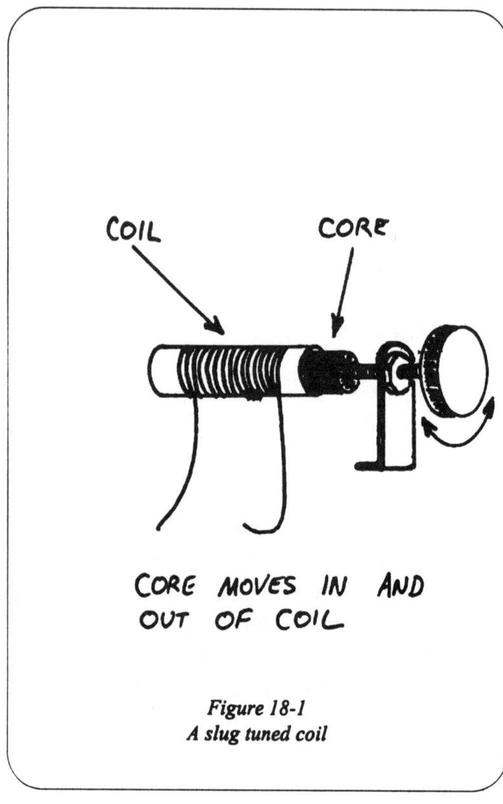

Figure 18-1
A slug tuned coil

is the mechanism to vary core material, windings, or dimensions.

Let's start with the core area. I have to admit that I draw a complete blank. How you would devise a coil to physically collapse or expand the diameter of it's windings is beyond me. Perhaps you might get away with a t r o m b o n e - s l i d e arrangement for a single-turn air-core coil in a high frequency circuit, but I don't see that as being adaptable to any broadcast band crystal set. I'll pass on this and leave it as fuel for one of your daydreams.

Next, we could try varying inductance of a coil by varying it's length. I have seen high frequency coils consisting of a dozen or less turns of comparatively thick wire. The coils were not wound on a form, but were rigid and self supporting. During calibration or adjustment of the circuit, a plastic screwdriver was used to spread or compress the windings in order to "tweak" the coil to the proper inductance value. This works well for minor one-time set-ups. It's not practical for circuits whose inductances must be continually adjusted.

On a purely conceptual basis, I can envision a spring-like coil made from brass. The spring would stretch horizontally, with a wooden dowel or glass rod inside of it, to support it. I assume that the coil would be insulated in some fashion, perhaps with nothing more than enamel. When the windings were compressed, the inductance of the coil would be maximized. When the coil was stretched and fanned out, the

Figure 18-2
A coil with insulation sanded off

inductance value would drop. I've never tried building the device I've just described, so I offer it only as an idea. For now, I'll skip trying to vary coil length.

Varying a coil's core material, on the other hand, is an excellent way to control it's inductance. The design begins with a cardboard or plastic tube, wound with an appropriate amount of wire. The coil has some nominal value, and could be considered an air-core coil. Next, a "slug" is inserted into the coil. A slug is a long, solid cylinder made of ferrite or similar magnetic material. It's mu is much higher than air.

As the slug enters the coil, air is displaced, and the high-mu core begins to influence coil. The inductance of the coil begins to rise. The greater the slug's penetration into the coil, the larger the coil's inductance. The inductance of the coil peaks when the slug is fully inserted.

Commercial slugs were sometimes manufactured with a thin, brass stem. The stem was threaded. By rotating the stem through a threaded bracket or support, it was possible to control the penetration of the slug very precisely. This allowed for very fine inductance

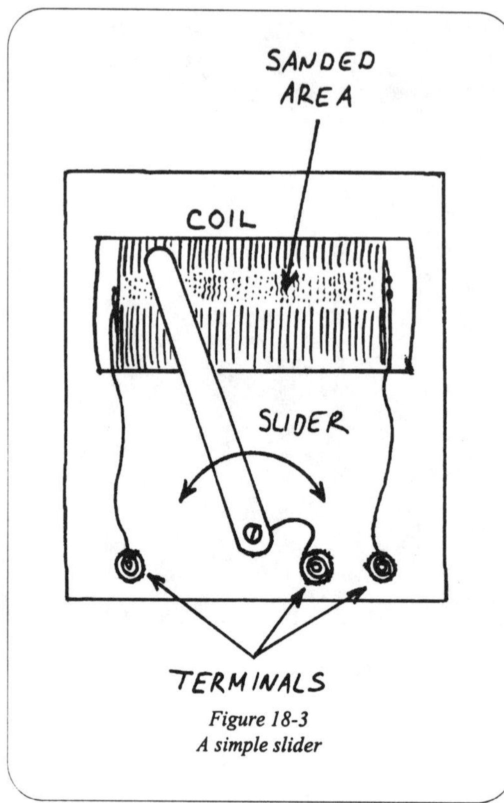

Figure 18-3
A simple slider

adjustments. (Figure 18-1)

For years, AM car radios were tuned using coils and slugs. The tuning knob as well as the pushbuttons were mechanically linked to a rack supporting several slugs. The rack inserted and withdrew slugs from several coils simultaneously. I imagine that this mechanism was judged to be more tolerant of vibration and contamination than tuners based on variable condensers.

The tapped coil, with which we began this chapter, allows you to "change" the number of turns of wire that contribute to the coil's inductance. As I indicated, the precision with which you can adjust the coil's inductance depends upon how many taps are used to divide up the coil. The more taps used, the smaller the incremental change from one tap to the next. It would be nice to have a tap for every turn of wire. By changing the design of our coil to use a slider, this degree of resolution is entirely possible.

At first glance, slider coils look like any other single layer coil. However, they are special in several ways. First, slider coils are best wound on rigid cores, with smooth, finished surfaces. The wire is wound on the forms with exceptional care, so that the resulting coil is neat, and orderly. No turns cross another, and there is no space between adjacent turns. Slider coils are most easily wound with magnet wire, and they favor the heavier gauges.

Once the coil is properly wound, fine sandpaper or emery cloth is used to wear away some of the insulation on the coil. Sanding is limited to a narrow strip that runs from one end of the coil to another. This must be done with care.

The object is to expose some copper from each turn of wire, without grinding too deeply. The insulation that protects each turn from adjacent turns must remain intact. (Figure 18-2)

The exposed windings are tapped with a slider. The simplest slider consists of nothing more than a narrow strip of springy metal. Brass is probably best, though aluminum or steel (from a can) have been known to work. A screw is passed through a hole at one end of the slider, and is used to anchor the slider to a wooden base. With the screw as a pivot, the slider is free to swing back and forth. (Figure 18-3)

The coil is mounted perpendicular to the slider, in such a way that the sanded wire region lies underneath it. The slider is in physical contact with the coil. As the slider is manipulated from side to side, it "slides" up and down the side of the coil. Wherever it contacts the coil, it creates a tap.

Because the slider is much wider than a single turn of wire, the slider is likely to touch more than one turn simultaneously. This can be avoided by soldering a short piece of bare copper wire to the face of the slider. The wire acts as a contact point, and should allow the slider to tap a single turn of wire on the coil. (Figure 18-4)

Coils are magnetic devices. They produce magnetic fields, and

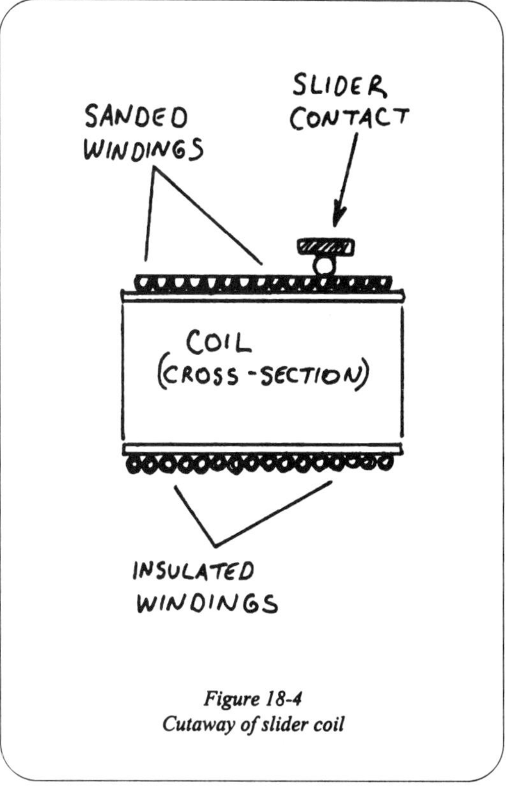

Figure 18-4
Cutaway of slider coil

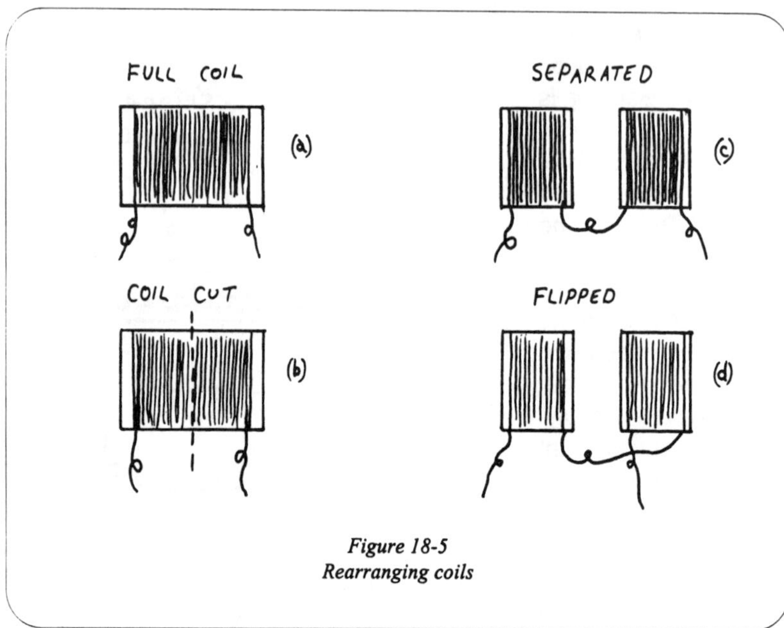

Figure 18-5
Rearranging coils

their inductive properties depend on those fields. While the equation doesn't explicitly say so, it can be inferred that, by altering those fields, we can alter a coil's inductance.

Imagine a simple, single layer coil, wound on a paper tube. The coil has some nominal inductance value. Let me describe a few modifications and I'll discuss their implications. (Figure 18-5)

First, let's cut the coil exactly in half. If I reconnect the severed windings with a piece of wire, and keep the two coil halves positioned exactly as they were, what happens to the total inductance? For all practical purposes, there should be no change. The magnetic field produced by each half-coil is free to influence the other half as it did before.

Next, let's draw the two coil halves apart. In this case, the total inductance will drop as the distance between the coil halves increases. The greater the distance between each half, the less influence each coil's field will exert upon the other.

A real interesting trick is to put the two halves back together, but flip one of them, 180 degrees. In theory, the inductive behavior of the coil should vanish! Why? Each half coil has it's own field, and fields have polarity. When the coils are combined and oriented properly, the

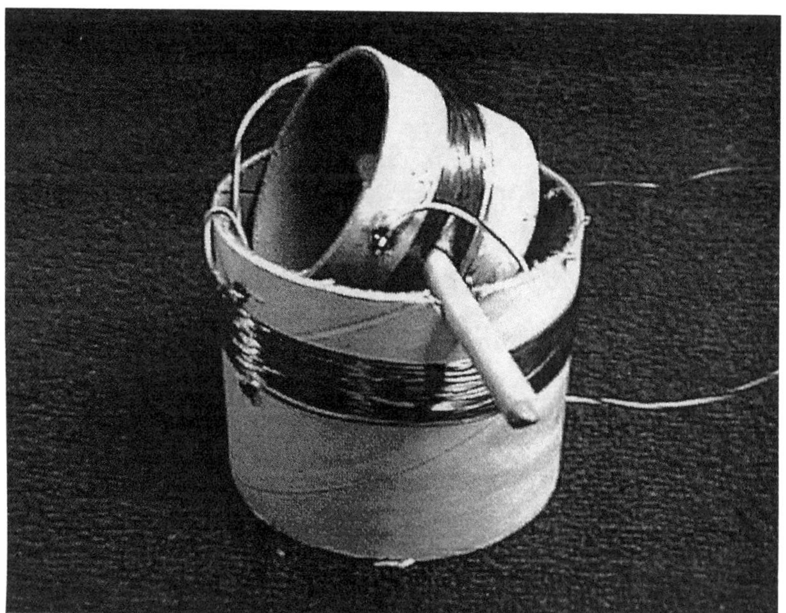

Figure 18-6
An experimental variometer

fields reenforce each other. If one half-coil is flipped around, it's field will be in opposition to the other coil's field. The cancellation of the magnetic field results in the cancellation of inductance.

A *variometer* is a type of variable inductor that uses this principle. A typical variometer consists of two coils, one large and one small. The smaller coil is mounted on a shaft, and suspended inside of the larger coil. The two coils are connected together in series so that their windings are all in the same direction, and their magnetic fields reenforce each other. In this position, inductance is maximized.

If the shaft is rotated 180 degrees, the small coil flips over, and it's magnetic field opposes the field of the large coil. Inductance is dramatically reduced. The shaft can be positioned at any angle 0 through 180 degrees. The angle between the coils determines the amount of reenforcement or opposition between the two coils, which in turn controls the inductance.

Figure 18-6 shows a simple variometer I put together using sections of two different mailing tubes. The control shaft for the inner coil is a 1/4" dowel. The magnet wire was salvaged from an electric

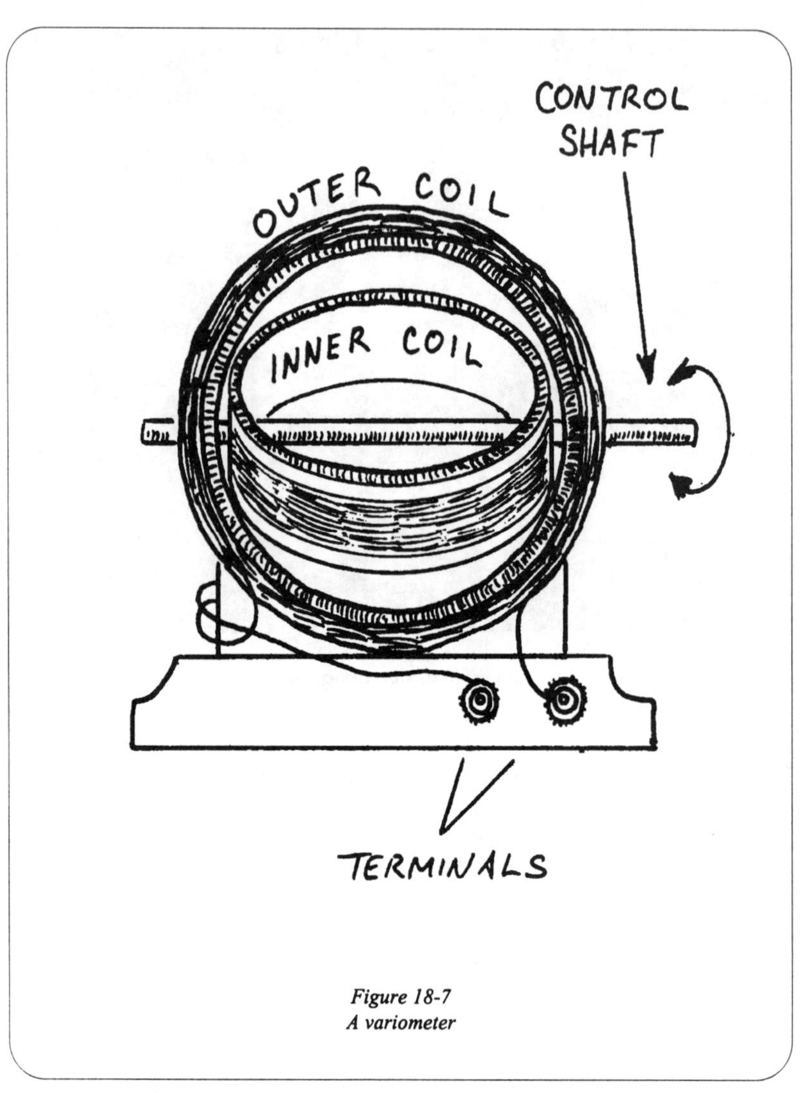

Figure 18-7
A variometer

valve. This experimental piece could easily be expanded into a nice laboratory piece. (Figure 18-7)

Chapter 19
The "Crank" Coil

There are a lot of ways to design and wind coils, and many in which to make them variable. Having played with slider coils quite a bit, I eventually began to think about how the basic slider design might be refined to produce a practical, stable, laboratory device. The resulting design, I call the crank coil. (Figure 19-1)

The crank coil is a practical implementation of the slider coil idea. It provides not one, but two fully adjustable taps, each capable of selecting individual turns. The slider mechanisms, based on screw threads, are manipulated with two small hand cranks (hence the name!) Slider positioning is very precise, and very repeatable.

I'd like to dive right into a discussion of the various components of this design, and show you how it's put together. Pay special attention to the illustrations. This device is not difficult to understand, it's just difficult to explain.

The heart of this instrument is the coil itself. The coil was wound on a mailing tube core, 2" in diameter and 5" in length. The first crank coil I wound used some pretty heavy magnet wire scavenged from an old motor. 180 turns of wire were laid down in a nice smooth layer. I left about a 1/4" of unwound tube at each end, and anchored the coil ends using my standard technique. (It may be helpful to refer to the chapter in which I discuss the winding of fixed coils.)

Since smoothness and uniformity in the winding process is so important for slider-based variable coils, I'd like to offer a tip for ensuring good results with salvaged magnet wire. Wire removed from old equipment may have kinks or waves in it. These defects have to be removed before a slider-type coil is wound. This is easily done by stretching the wire. (Figure 19-2)

Take your salvaged wire, and tie one end to a doorknob or other stable anchor. Walk away from the door, and pace off as much wire as you think you'll need. Examine the wire, taking care to manually remove any knots or little loops. Return to the free end of the wire, and tie it to a large screwdriver.

Now, using the screwdriver as a handle, place the wire under tension. Gently increase the tension until you feel the wire just begin to "give". Don't over do it, or you'll risk breaking the wire. Copper is very soft, and will stretch under tension. The diameter will be reduced only slightly, but the end result is that most of the annoying kinks and bends in the wire will magically disappear.

Once the my coil was complete, I devised a mounting scheme for it. The coil was supported at it's ends by two end-blocks. The end-blocks, in turn, were attached to a 5" by 7" pine base. The base, which features a fine routed edge, was purchased at a local craft shop.

Each end-block was 2 ½" tall and 3 3/4" wide. Using a compass, a 2" circle was inscribed on the inside face of each end-block. With the circle as a guide, I machined a circular groove into each end-block. The groove can be cut with a fine wood chisel, or a sharp knife, but a small router bit in a hand-held grinding tool proved to be ideal. As an alternative, you might try using a 2" hole saw. In any event, the grooves were cut just deep enough to accept the ends of the coil. See figure 19-3.

Each end-block was drilled to accept a connection terminal composed of a 6-32 screw, washer, nut, and wing nut. A short wire was soldered from the anchors at each end of the coil to the terminal screws. The soldering was done from the inside of the coil. The end result, to this point, was a nice, fixed value inductor. (Figure 19-4)

Next, a slider mechanism was installed on each side of the coil. Each mechanism consists of a slider block with a spring contact, a threaded crank rod with crank, and a guide tube. For the sake of clarity, I will describe the mechanics of one of the slider devices only. Remember that the components in the following text are duplicated on the other side of the coil as well.

The slider block is a small

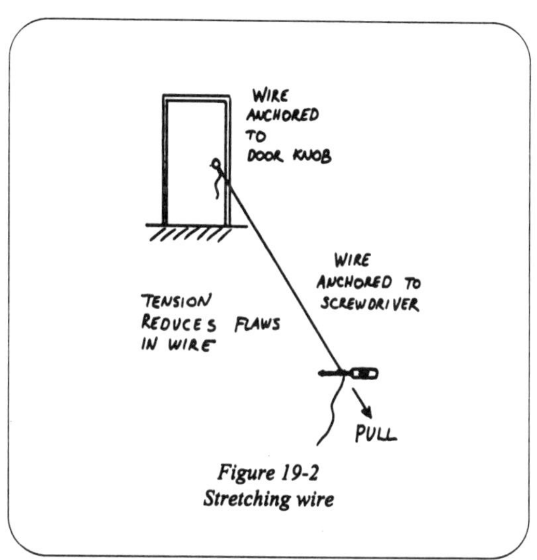

Figure 19-2
Stretching wire

Figure 19-3
End block detail

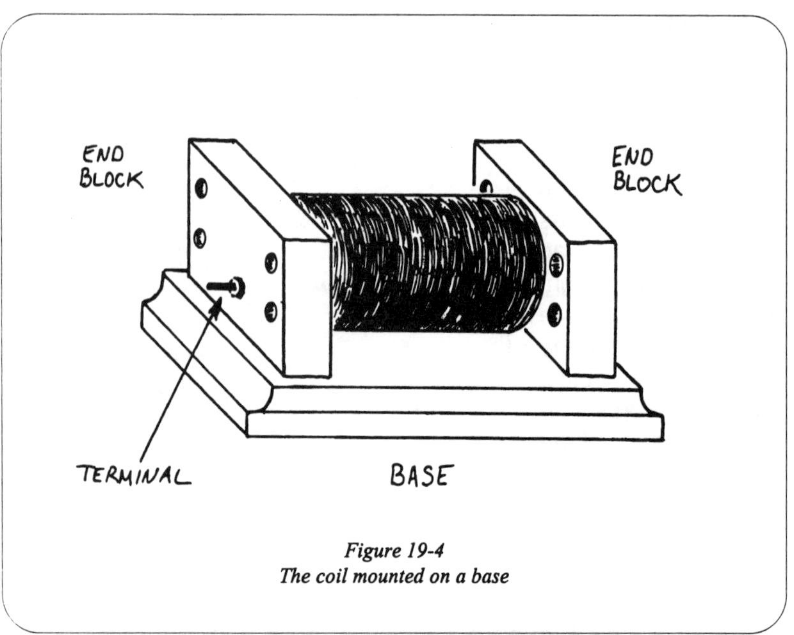

Figure 19-4
The coil mounted on a base

rectangle of dense, insulating material, 1 ½" tall, ½" wide, and 3/4" deep. My slider blocks were fashioned from a scrap of phenolic plastic, though other plastics or any dense, hardwood would work as well.

Two holes were drilled in the block. The top hole measured just under 1/4". A tap was used to cut threads into it. The bottom hole measured about 3/16". A short piece of brass tubing was forced into the hole to create a smooth, low-friction bushing. (Figure 19-5)

The end-blocks were drilled to admit the crank rod and the guide tube. The crank rod was cut from 1/4" threaded brass rod. The guide tube was cut from a length of thin-wall brass tubing, selected to fit into the bushing previously pressed into the slider block. Properly installed, the crank rod and the guide tube are precisely parallel to each other, and to the axis of the coil. They were spaced from each other, about 3/4".

When the slider mechanism is assembled, the threaded rod passes through the top hole, and the guide tube passes through the bottom. Rotating the crank rod clockwise causes the block to screw it's way toward the front of the coil. Counter-clockwise motion screws the slider block toward the rear of the coil. (Figure 19-6)

Some comments about the crank rod are in order. The rod was

fitted at one end with a small crank, which is used to rotate the rod. My cranks were made by hand from bits of brass scrap and other odds and ends. You could use large Bakelite or plastic knobs.

The other end of the rod was secured with two brass nuts. Using two wrenches, one nut was tightened against the other to lock or "jam" them together. The jam nuts must be adjusted to allow easy rotation of the rod, with minimal side-to-side motion of the rod.

I soon learned that the holes drilled through the wooden end-blocks make lousy bearings. My solution was to line each hole with a short segment of brass tubing to act as a bushing. Because the finished

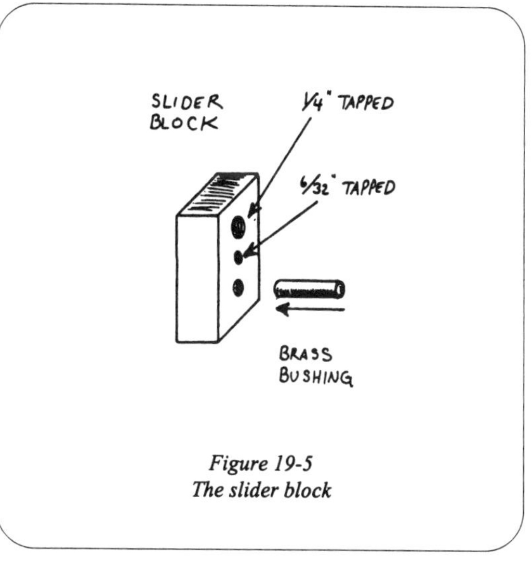

Figure 19-5
The slider block

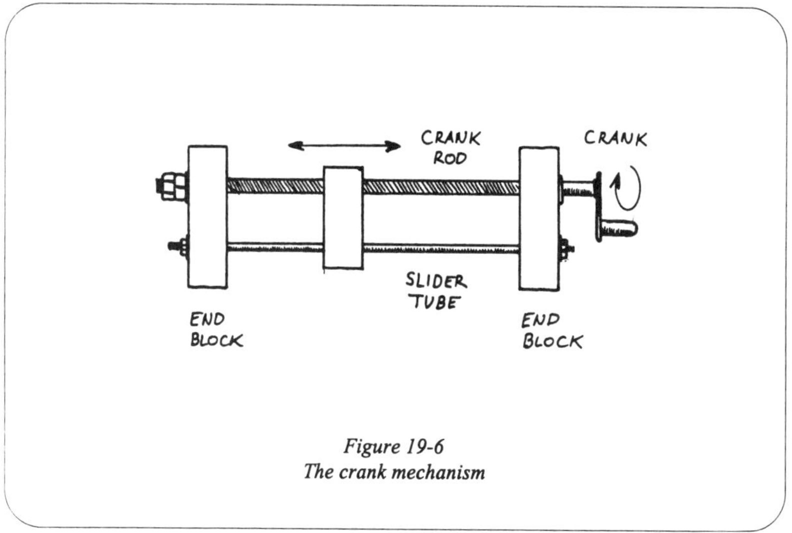

Figure 19-6
The crank mechanism

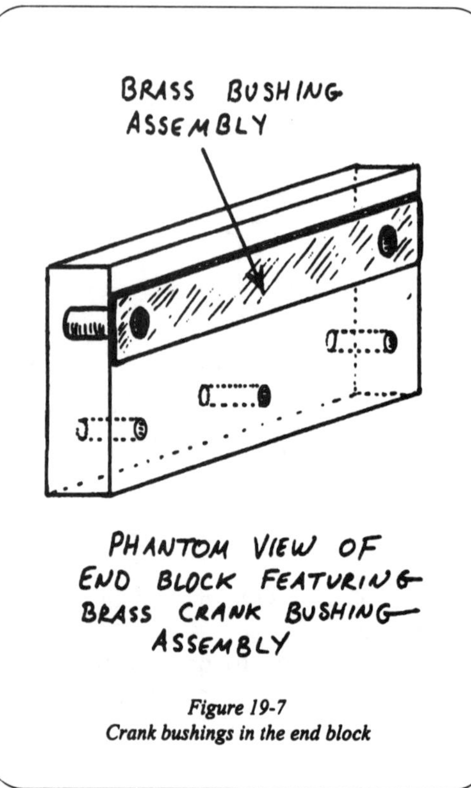

Figure 19-7
Crank bushings in the end block

coil has two crank rods, each end-block needs two bushings. I fashioned a clever two-bushing assembly by soldering both of them to a ½ " strip of heavy brass stock. The strip supports, aligns, and holds the bushings in place. The strip itself was anchored to the end-block with two small, brass nails. (Figure 19-7)

The guide tube is roughly 6" long. Actually, it was trimmed so that it's ends were flush with the outer faces of the end-blocks. To keep the guide tube in place, I inserted a 7" length of 6-32 threaded brass rod. On the front end-block, the rod was terminated with a washer and nut. At the rear block, the rod was also terminated with a washer and nut, but was allowed to extend about 3/4". A wing nut was added. As you will see, this provides a terminal and binding post for the tap. (Figure 19-8)

Let's take inventory. At this point, I had a carefully wound coil with wing nut terminals mounted to stable base using wood end-blocks. I had installed a threaded crank rod, a guide tube, and a slider block. I fiddled with the alignment and mechanical details until the slider block would move smoothly from end of the coil to the other. A 1/8" gap was established between the face of the slider block and the coil. I double-checked to make sure that the gap remained constant over the travel range of the slider block. The only part missing was the spring contact. Figure 19-9 shows the crank end of the coil assembly.

A suitable contact has several crucial attributes. It must be

Figure 19-8
Terminals

springy enough to maintain good electrical contact with the coil. The contact point itself has to be narrow enough to select a single turn of wire on the coil. On the other hand, copper wire is a soft material, so a sharp, pointy contact, would eventually tear into the side of the coil.

The ideal contact turned out to be a common B-B. B-B's are copper plated, making them good conductors. They are small enough to be able to contact single turns on the coil, yet their smooth shape prevents damage to the wire. A springy strip of brass stock was selected. A hole was drilled near one end of the strip, small enough to allow a B-B to rest in the hole without falling through. The B-B was carefully soldered to the strip from the back, to avoid contaminating the front surface of the B-B.

The strip was wrapped halfway around the slider block, and assumed the shape of a squared-off "U". One leg of the "U" held the B-B in firm contact with the coil. A 1" piece of insulated wire was soldered to the other end of the "U". A hole was drill through the base of the "U" into the side of the slider block. The slider block was threaded with a 6-32 tap, and the spring contact was fastened to the slider block with a 6-

Figure 19-9
The cranks

32 pan head screw. (Figure 19-10)

The final step was to drill a hole in the bottom of the slider block, just deep enough to reach the brass bushing in the guide hole. The short wire dangling from the spring contact was inserted into hole at the bottom of the slider block, and soldered to the bushing. (Figure 19-11)

Just to clarify the electrical circuit, let me summarize the operation of the coil. When the crank is turned, the crank rod rotates, which screws the slider block to the left or to the right. The B-B, which is pressed against the coil by a brass strip, establishes an electrical tap on the coil. The signal flows from the coil tap, through the B-B, through the brass strip, and through the 1" wire to the brass bushing in the slider block.

The bushing passes the signal on to the guide tube, which in turn conducts the signal to the 6-32 wing nut terminal at it's rear. The guide tube actually does double-duty. It not only guides the motion of the slider block, but it provides a movable electrical connection to the tap.

You may have noticed that I failed to mention sanding the coil to expose the wire. Let me assure you that there is method to my madness, and my answer is that you *don't* sand the wire....until the coil and slider is completely finished!

Yes, I could have sanded two narrow contact areas when I had finished winding the coil.

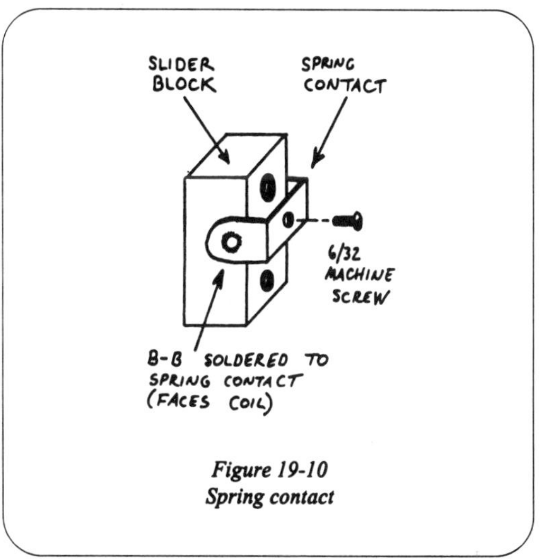

Figure 19-10
Spring contact

Unfortunately, the sanded areas of the coil must be perfectly straight. The sanded region on one side of the coil must be perfectly parallel to the sanded region on the other side. This sort of precision is difficult to achieve simply by "eyeballing" the job.

On the other hand, if you wait until the coil is completely finished, all you need to do is work the cranks. If you send the slider blocks back and forth a few times, you'll discover that the B-B scratches the insulation on the wire. I wouldn't rely on this to make good electrical contact, but if you then take the coil apart, the scratches will provide accurate guidance for your sanding efforts. This way, you know precisely where to sand some insulation off the coil, because you know precisely where the B-B contact will track.

Of course, once the coil and all it's components were adjusted and working properly, I took it completely apart. All of the brass parts were cleaned with brass polish. The wood components were stained and coated with polyurethane. The device was then reassembled. Green felt was glued to the bottom of the base for a touch of class.

End Results and Going Farther

I've been very happy with this coil. Depending on the circuit used with it, the coil seems to be capable of some fairly sharp tuning.

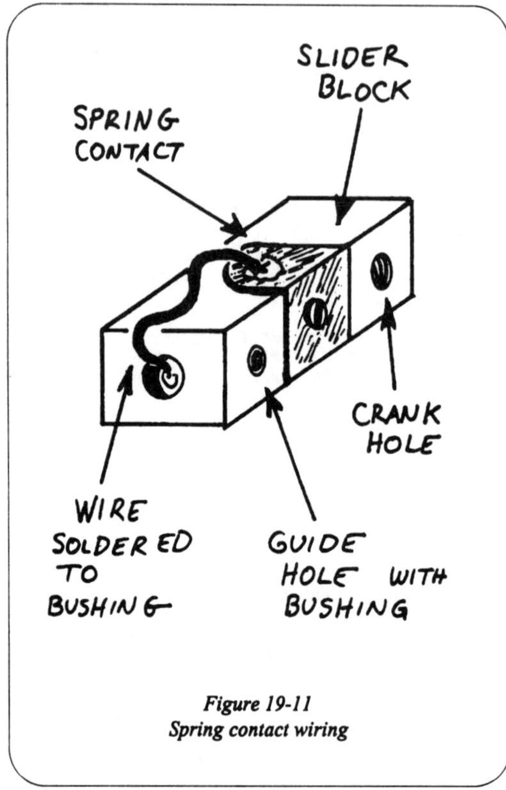

Figure 19-11
Spring contact wiring

Adjustments are quick, easy, and precise. I can set either slider, move it, and then return it to the initial position simply by counting rotations of the crank.

Another attractive feature of the crank coil design is the relative ease with which coils can be changed. Since building the crank coil, I have wound chunks of mailing tube with other gauges of wire. By simply removing one of the end-blocks, I am able to remove the existing coil and install and experiment with alternates.

One warning is in order regarding cardboard mailing tubes as coil forms. My first crank coil was wound, tested, and then set upon a shelf while I tinkered with the other components in this book. At some point, a few months later, I was dismayed to discover the windings on the coil very loose and sloppy. The slack in the windings was so great that the spring contact actually snagged on the wire, tearing the coil apart. What in the world had happened?

After investigating several theories, I finally concluded that the cardboard tube had *shrunk*! The humidity in Tucson, Arizona, where I live, is often very low. As the tube aged and dried out, it simply contracted.

The solution is to coat the entire coil, inside and out, with polyurethane or varnish. Not only will this preserve the moisture content in the cardboard, it provides additional mechanical support to the individual windings.

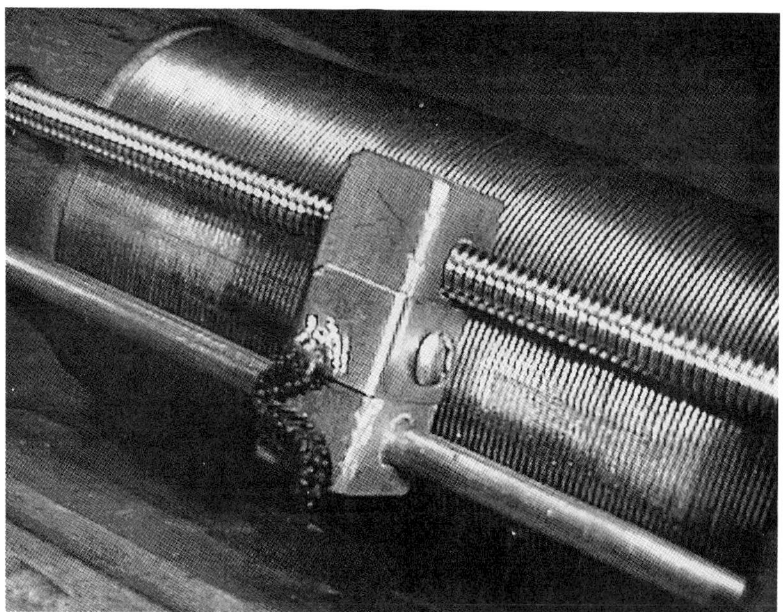

Figure 19-1
The "crank" coil

Several weeks after the coil was completed, I was fortunate to find a copy of *Hawkins Electrical Guide*, circa 1927. I was delighted to find engravings that depicted tuning coils very similar to mine. One illustration featured a tuning coil with "...two slides and is used where closer tuning is required than can be obtained with a single contact coil." I actually felt a tinge of pride to discover that Marconi tuning coils actually employed small balls as the preferred method to contact slider coils!

Isn't this a remarkable coincidence? No, absolutely not! This merely demonstrates that analyzing a problem's solution in terms of essential characteristics will yield results. I tried to design an effective slider coil. I had no specific knowledge of vintage designs, but I had identified the essential characteristics of a suitable solution.

Seventy sum-odd years after the fact, I recreated a design that, in it's day, was state-of-the-art. Imagine the potential benefit from applying the "essential characteristic" approach to modern problems...

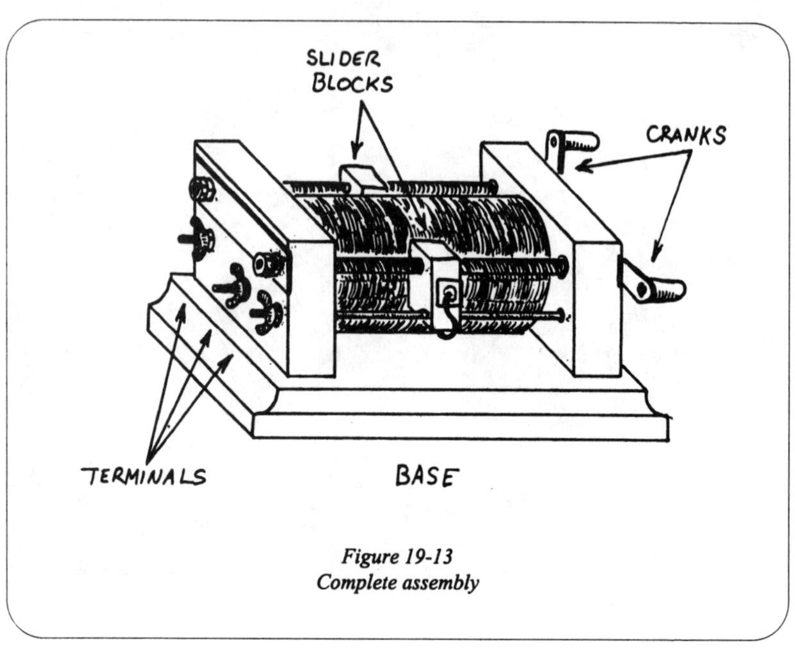

Figure 19-13
Complete assembly

Chapter 20
Thoughts On Simple Tuners

everal chapters back, I introduced the common tuning fork as a vehicle for describing the transmission and reception of radio waves. A tuning fork that's energized with a good whack will ring, and broadcast sound waves with a very specific frequency.

More important, you learned that a silent fork, when stimulated by a tone whose frequency matches it's own, can capture enough of the broadcasted energy to ring all by itself. If the fork is exposed to sounds that don't match it's own frequency, no energy transfer takes place, and the fork remains silent. The ability to capture and concentrate energy from waves of the proper frequency, while ignoring all others is a property called resonance.

I went on to say that, together, a coil and a condenser can do for radio waves what a tuning fork does for sound. Having spent a good many pages describing theory, characteristics, and construction tips for both condensers and coils, it's time to combine them to produce a simple electronic tuner. You might say that this chapter is where the rubber meets the road.

In the case of the tuning fork, it's resonant frequency depends on the physical characteristics of the fork. These include the mass and length of the tines, the material of which they're made, and other factors. In the case of the simple tuners I will present here, frequency depends upon the values of the condenser and coil from which they're built.

Figure 20-1 shows the basic formula for electronic resonance. "L" represents the inductance value of the tuning coil in henries, and "C" is the value of the tuning condenser in farads. The Greek character "pi"

$$f = \frac{1}{2\pi\sqrt{LC}}$$

Figure 20-1
The formula frequency of tuned circuit

is a mathematical constant (about 3.141). "f", of course, is the resulting frequency in hertz, or cycles per second.

Let's toss some numbers into this formula to get an idea of how the circuit would behave. Suppose that I planned to build a tuner using a coil with an inductance of 1 micro henry, and a condenser with a capacitance value of 1 microfarad. Plugging these values into the formula, I compute that the resulting tuner circuit should resonate at 159 kilohertz (159,000 cycles per second).

You can also work the formula backward. If you have a desired frequency, you can plug it in and solve for the condenser and coil that you'll need.

A drawback to building a tuner with a fixed coil and a fixed condenser is that the resonant frequency is also fixed. On the other hand, if we make either the capacitance or inductance value adjustable, we can then alter the resonant frequency. Adjustable condensers and coils allow you to easily change the tuner's resonance value to any of a thousand different frequencies.

We can build an adjustable tuner with a fixed condenser and a variable coil. Or, we can use a fixed coil and a variable condenser. If you want ultimate flexibility, nothing beats combining an adjustable coil of some type with a variable condenser. In any case, the resonant frequency of the circuit will vary according to the formula.

How do you wire up a tuner? There are two basic ways, each of which reacts differently to resonant conditions.

The first, is a series resonant circuit. Figure 20-1 shows a very simple radio receiver based on the series resonant tuner. Signals received by the antenna are fed to the tuner. Only signals that match the resonant frequency of the tuner are allowed to pass through to the detector and headphones. By adjusting the condenser or the coil, you can change the resonant frequency of the tuner. This changes which signals are allowed to pass to the detector and headphone, and thereby "tuning" the radio.

The other way to wire a tuner circuit is to connect the coil and condenser in parallel. Figure 20-3 shows a typical hookup for a radio using a parallel resonant tuner. Signals are gathered by the antenna and fed into the tuner. Parallel tuners are good conductors for non-resonant signals, so they pass through the tuner and are dumped into the ground connection. Those signals never make it through the detector or into the headphone. On the other hand, parallel tuners will block signals at resonance, so any signal that matches the resonant frequency of the tuner will be forced to pass through the detector and headphones.

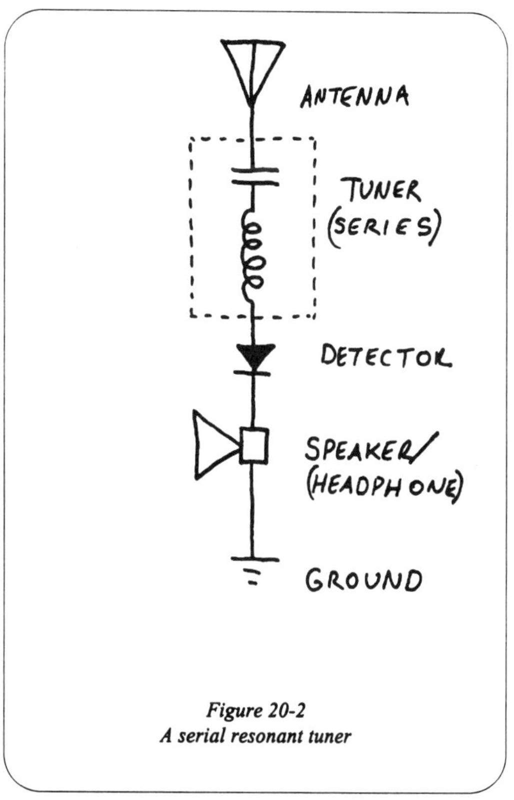

Figure 20-2
A serial resonant tuner

An ideal tuner is very picky. It will discard all signals except those that match it's resonant frequency, exactly. Unfortunately, ideal tuners don't exist, particularly where simple crystal set designs are concerned. In the real world, tuners are much less discriminating, and can actually be quite sloppy. They "prefer" signals at the resonant frequency, but will accept other signals if their frequency is close enough to the tuner's resonant frequency. The amount of rejection varies with frequency, and it increases as the applied signal deviates more and more from the resonance value. Real-world tuners actually accept a range or span of frequencies, centered around the ideal value.

Tuners are sometimes described in terms of their *selectivity*. The selectivity of a tuner is the precision with which it can tune a

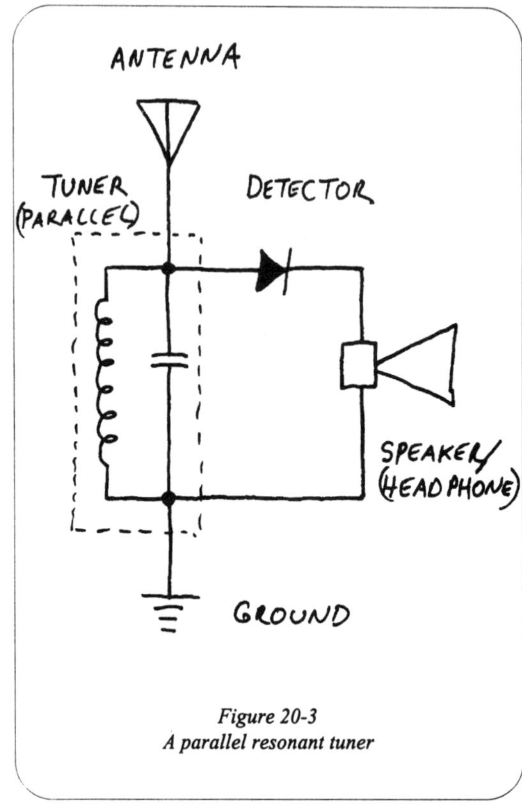

Figure 20-3
A parallel resonant tuner

specific frequency. A highly selective tuner will accept signals at or very close to it's resonant frequency. A tuner with poor selectivity will tend to allow a wider assortment of signals through, even those that differ significantly from the resonant frequency.

Selectivity becomes a practical concern when trying to select or tune one of two stations that are adjacent on the dial. A tuner with good selectivity has a narrow span, and will accept energy only from the desired station. A tuner with poor selectivity will allow energy from adjacent stations to bleed into the desired signal. You may hear two or more stations in your headphones at once! (Figure 20-4)

If you find that your tuner exhibits poor selectivity, the fault may lie in any of a number of places. Condensers and coils are devices whose performance can be degraded by a poor choice of materials or construction technique. Components that are "lossy" or wasteful of energy will form inferior tuners. Review the essential characteristics of these components and reconsider your design.

The other concern is a factor called loading. If we start with a bright, shiny, 440 Hz tuning fork, and decide to paint it, we may discover that we've ruined it! The painted fork will not ring as loudly as it once did, and it's frequency may actually be altered. Why? The paint adds mass to the fork, and in subtle ways, restricts the movement of the tines. The added work, or load, associated with moving the painted tines causes changes in the resonant behavior of the fork.

The same effect is visible in resonant circuits like tuners. In simple crystal sets, signals from the tuning circuit are fed to the detector and the headphone. It takes work to overcome the losses in the detector, and it takes work to drive the headphone hard enough to produce sound. The tuner, which has no power source of it's own, must capture and concentrate feeble radio waves, and then muster the energy to drive the set. If the energy requirements of the detector and headphone represent a significant load to the tuner, tuner performance will suffer.

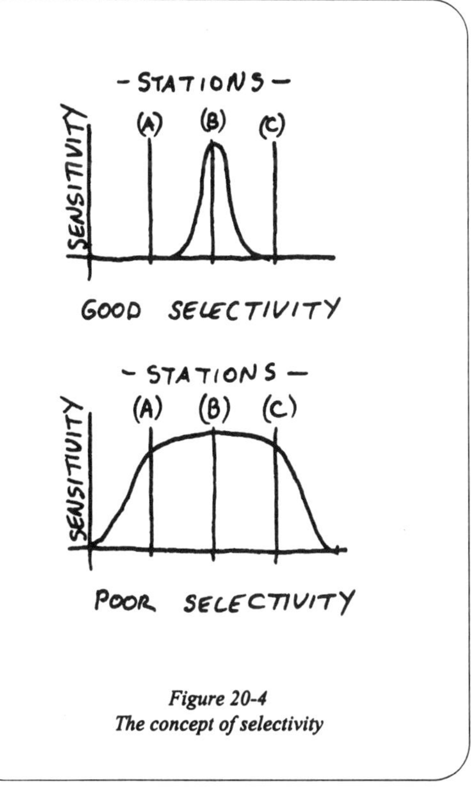

Figure 20-4
The concept of selectivity

The goal, then, is keep loading to a minimum. As I mentioned in the chapter on headphones, one way to help minimize loading is to increase the impedance of the headphone. Another approach is to modify the way in which energy from the tuner is coupled to the detector and headphone.

We know that a coil, if energized, will produce a magnetic field. We also know that if the coil is de-energized, the field cannot maintain itself and will collapse. As the field vanishes, the energy that was stored in it reappears in the wire as electricity.

If we wind a second coil of wire on the same tube, the magnetic field from the first coil can influence and even generate electric currents in the second coil. Purely through magnetic means, it's entirely possible to transfer energy from one coil to the other. This two-coil device is

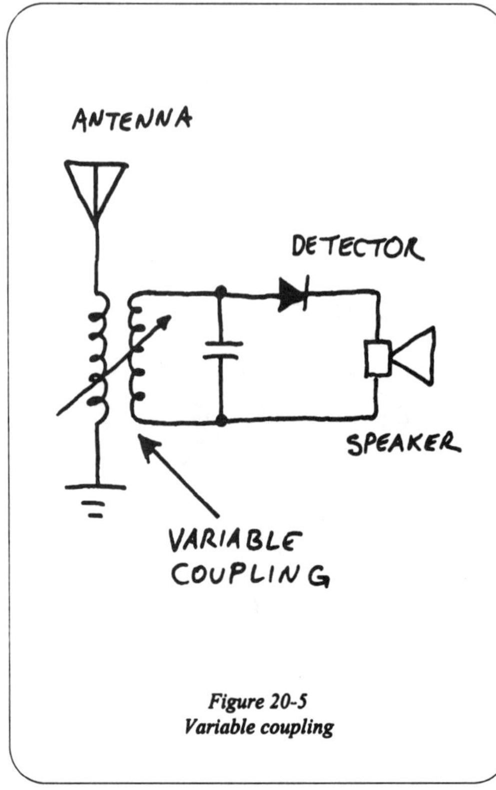

Figure 20-5
Variable coupling

called a *transformer*.

If the transformer coils are in very close proximity, the transfer of energy from one coil to the other can be quite efficient. The coils are said to be closely coupled. On the other hand, if we separate the coils, the transfer efficiency will degrade as the distance increases. The coupling becomes loose.

The orientation of the coils is also important for coupling. If the two coils lie end to end, coupling will tend to be very tight. If one of the coils is rotated 90 degrees, so that the axis of the two coils are at right angles to each other, coupling will be very poor.

Do you recall an adjustable inductor design called a variometer? It consisted of two coils, connected together, with one inside the other. The inside coil was mounted on a shaft so that it's angular relationship to the outer coil could be changed. The inductance of the variometer changed as the shaft was rotated. If we cut the link, and treat the inside and outside coils as two separate windings, we have transformer whose coupling can be changed by rotating the shaft!

One way to control tuner loading is to insert a transformer between the tuner and the detector/headphone. By varying the distance or the angular relationship between the transformer coils, we vary the degree of coupling. If we decrease the coupling, we decrease the amount of energy that is transferred from the tuner to the detector. As a result, we reduce the amount of loading. See figure 20-5

So, loose coupling with a transformer is a good thing, because it reduces loading on the tuner and improves selectivity, right? Yes and no.

Loose coupling will improve the selectivity of the tuner. However, loose coupling means a reduced transfer of energy from the tuner to the detector. The result is lower headphone volume. The receiver may prove to be less sensitive to very weak signals.

In crystal sets, good selectivity and good sensitivity tend to be mutually

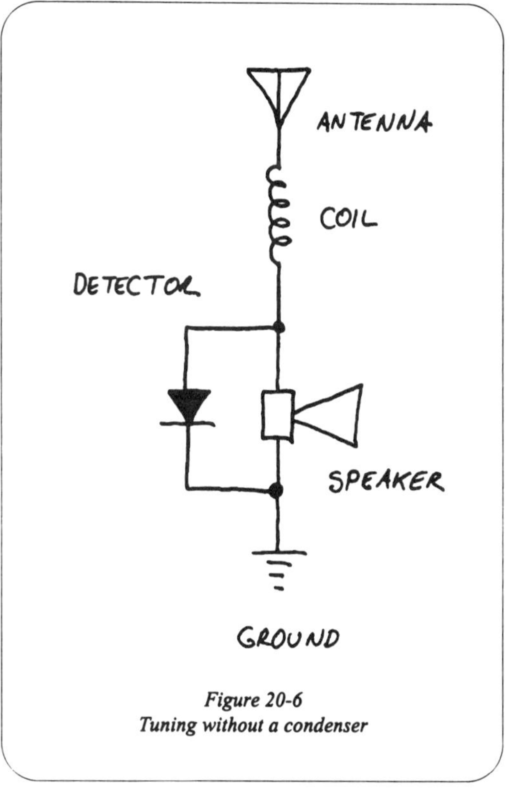

Figure 20-6
Tuning without a condenser

exclusive goals. Optimizing a set for one tends to degrade it's performance in regard to the other. The good news is that this is not as much of a problem as you might guess. The trick is to optimize the crystal set for it's particular application.

For example, If you live in a large city with lots of powerful, local radio stations, selectivity is probably more important than sensitivity. You want your tuner to be able to discriminate between one station and another. Conversely, if your goal is to hear weak and distant stations, you'll want to optimize the set for sensitivity and high volume. Selectivity suffers, but may not be a problem for the stations you're interested in.

Figure 20-6 should be thought provoking. Notice that there is no condenser! How can you have a tuner without a condenser! The answer is that there *is* a condenser...it's just hidden. We defined a condenser as any pair of adjacent conductors separated by an insulator. Well, every winding on the coil has a neighbor that runs adjacent to it. Each

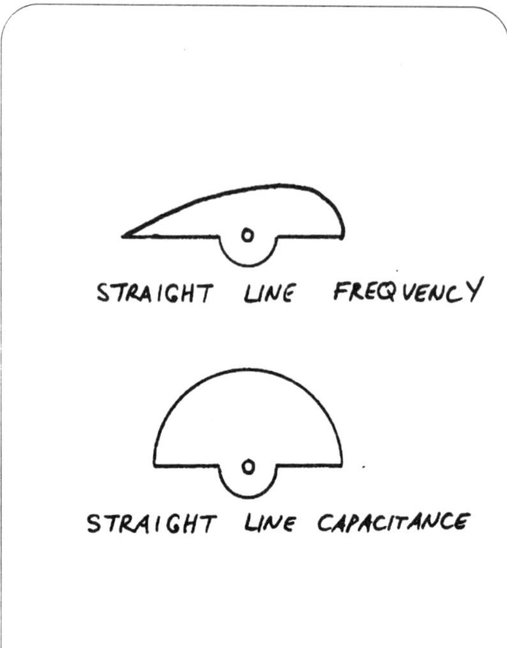

Figure 20-7
Condenser blades:
straight capacitance vs straight frequency

pair of windings acts like a very tiny condenser. Depending on the type of coil and the way it's been wound, this hidden capacitance can be significant.

The hidden capacitance in the coil can cause problems for you when you attempt to build tuners for higher frequencies. The formula will tell you that higher frequencies require smaller inductance and capacitance values. If the formula says that you need a condenser that's smaller than the capacitance hidden in the coil, you can't use the coil! Even if you run with no external condenser, it's resonant frequency will always be lower than the value you want. You're only choice would be to redesign your coil. Low-capacitance coils, like the basket weave design, might be an option.

During your experiments, you will stumble upon other hidden condensers. The sky and the ground, both of which tend to be conductors form an enormous condenser. Waving your hand near a crystal set may change it's tuning, because your hand acts as the plate of yet another hidden condenser.

In some cases, it is the inductor that's hidden. Remember that all conductors have a certain amount of inductive behavior. Tubular condensers with rolled plates may have significant hidden inductance. It's not difficult to see the similarity between the multiple turns in a coil, and the layers of a wound condenser.

When your tuner uses a shaft-adjustable variable condenser (like the roofing metal condenser), a knob and a dial is a nice touch. After you snag your first radio station, you can use a pencil to mark the dial position. That way you can return to that setting later. The dial becomes even more useful as you add more and more stations over time.

You will notice that, unlike a modern tuner, stations are not evenly spread across the scale. In fact you'll find that the stations will tend to crowd together near one end of the dial. Why is that?

Imagine a wheel, with a circumference of two feet. If we place the wheel on the ground, and roll it exactly one revolution, the wheel will have traveled two feet. After two revolutions, the wheel has moved four feet. After ten revolutions, the wheel has moved twenty feet, and so on. The point is that there is a direct and *linear* relationship between revolutions and distance traveled. For every incremental change in revolution, there is a finite, incremental change in distance traveled.

The roofing metal condenser described in a previous chapter used rotor blades that are semi-circular. As the control shaft is rotated, more and more of the rotor blades are shaded by the stator blades, and the capacitance increases. This is also a linear relationship. Move the shaft 10%, and the capacitance will increase 10%. A variable condenser of this type is called a straight-line capacitance condenser.

Now, take the frequency formula and set the L term to some fixed value (it doesn't matter what, just assume you are using some sort of fixed-value coil). Make a list of ten evenly spaced capacitance values, and plug them into the formula. What you will discover is that the frequencies computed will *not* be evenly spaced. Because the L and C terms are not only in the denominator of the fraction, but are under a square root sign as well, the resulting frequencies *cannot* be linearly proportionate to C. An incremental change in C won't always produce the same incremental change in frequency. Frequency is not linearly proportional to capacitance.

This means that stations on a simple tuner won't be evenly spread across the dial, despite the fact that the condenser produces evenly-spaced capacitance values. This accounts for the "crowding" affect I mentioned a few paragraphs ago.

How can you make allowances for this effect? How can you produce a simple tuner that tunes frequencies at regular interval? What you need is a *straight-line frequency condenser*. In the case of the roofing metal condenser, the straight-line frequency version is built just like the straight-line capacitance version. The difference lies in the shape of the rotor blades.

Instead of semi-circular plates which produce linear changes in capacitance as a function of shaft rotation, the modified plates are purposefully non-linear. The shape of the blades generates an odd sequence of capacitance values as the shaft is turned, engineered to cancel the effects of the square-root and the inversion. The result is that tuner frequency becomes linear as a function of condenser shaft rotation.

The plates, which have a very definite profile, remind me of the back of a whale as it begins to surface. One of the nice things about the roofing metal condenser, is that you can easily fit it with different plate shapes to accommodate linear tuning. See figure 20-7.

Mathematically speaking, frequency is also non-linear with respect to L for all the same reasons. If you design some sort of variable coil that you plan to calibrate with a dial, understand this: Either the coil will have to be designed to counteract the square root in the frequency formula, or expect to have stations spread unevenly across your dial.

Chapter 21
Some Thoughts On Antennas And Grounds

A pinwheel is a completely passive energy collection machine. It has no internal energy source. In the absence of wind, it can do nothing. A slight breeze, however, is captured by it's blades, and it will begin rotating with hesitation. As the speed of the wind picks up, the pinwheel is able to harness more energy. The rotation of it's blades becomes quick and sure. Yet, the motion of the pinwheel can be no more energetic than the wind that strikes it.

Like the pinwheel, the crystal radio is a passive device, and depends upon external energy as well. It's performance hinges on the strength of the signal it captures. The power of the sound produced in it's headphones can never be greater than the power of the signal received. Because of the laws of physics, it will always be much less. So, it pays to make the radio as efficient a "wave catcher" as possible.

Antennas and grounds are to the crystal radio what blades are to the pinwheel. Both capture energy from the machines' environment. The more efficient the capture mechanism, the more energy that's available to drive the machine.

Real antenna design is a subject so broad and deep that it has occupied many careers if not a lifetimes. My favorite radio books seem to fall into two camps: those that cover the material with all the excitement of a metro telephone directory, and those that gloss over the subject entirely. I suspect that the latter authors are as mystified by these excruciating details as I am. Luckily, a few practical rules of thumb will give the casual experimenter acceptable results.

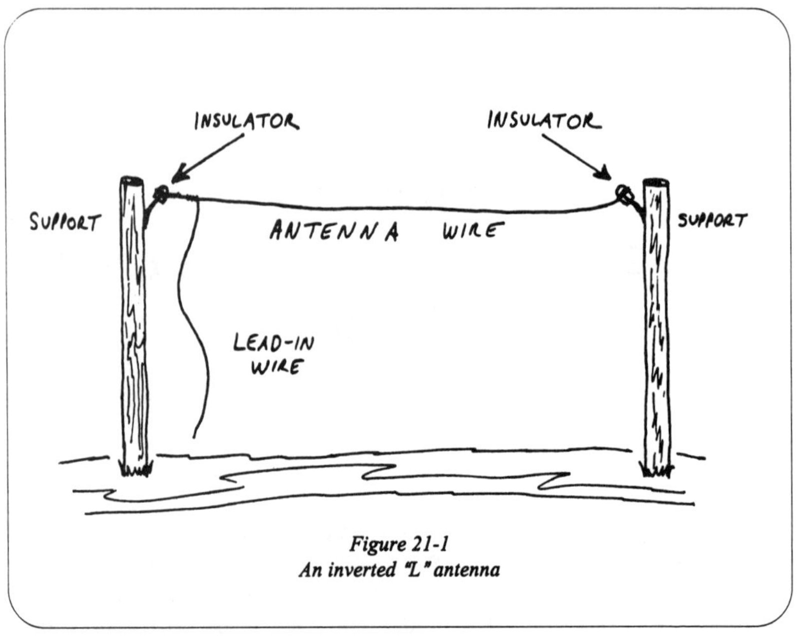

Figure 21-1
An inverted "L" antenna

A classic crystal radio antenna is a long piece of wire, suspended in the air, stretched between two supports. The higher the wire is from the ground, the better. Length helps, too. I've seen references to antenna wires in the fifty to one hundred foot length range. Wires are typically stretched between two tall trees, a house and a tree, or between wooden poles. (Figure 21-1)

For best results, the ends of the wire must be insulated from it's supports. Old telegraph insulators work well, as do the necks of broken wine bottles. You can buy ceramic egg-shaped insulators that allow you to couple the antenna wire with a piece of cord or anchor wire without the two coming into contact. (Figure 21-2)

In a pinch, you can use just about any type of wire you please. Be aware, however, that copper is a soft material, and a long length of copper wire may stretch and even break under it's own weight. You might try steel wire by itself, or a strand of steel wire supporting the copper wire.

Trees and poles tend to sway in the wind, so you'll want to leave some slack in the wire. An alternative is to use the egg-type insulators, and anchor the antenna to it's supports with some old screen door springs. That way, the antenna can "give" without breaking.

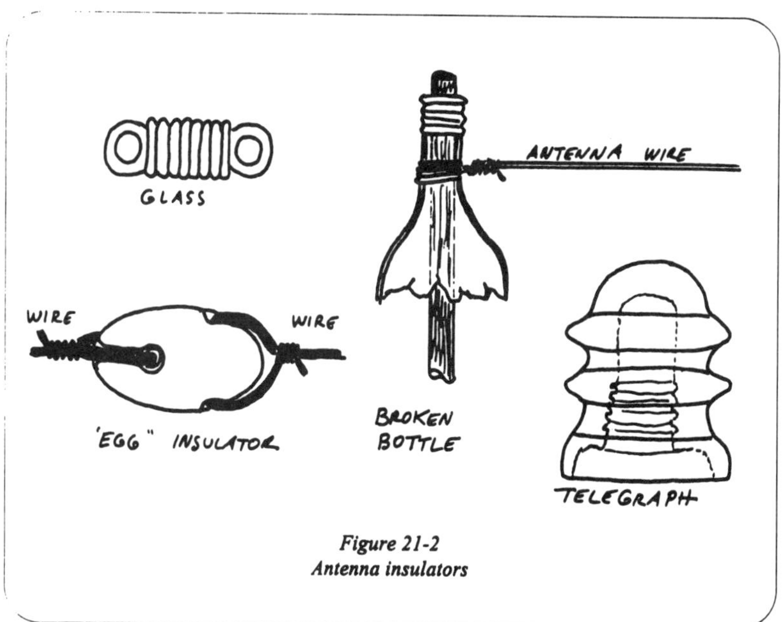

Figure 21-2
Antenna insulators

(Figure 21-3)

To get the signal down to your receiver, you need to attach a "lead-in" wire. This is simply another piece of copper wire soldered to that antenna wire near one of it's ends. The lead-in should be insulated, and prevented from touching walls or any other supports. This design is often referred to as an "inverted L" antenna. If you consider the relationship between the antenna and lead-in wire, the reason for the name becomes obvious. Some electronic stores sell complete kits containing antenna wire, insulators, and accessories.

There are several things to keep in mind if you are contemplating setting up an inverted "L". First, **NEVER** install this or any other antenna over, under, or near power lines. To be blunt, sooner or later somebody will end up dead. Assuming your antenna is a safe distance from any power line, try to orient the antenna perpendicular to them. This helps minimize static caused by the power lines. Finally, any external antenna is subject to a lightening strike. You will want to purchase and install a lightning arrester. I'll talk about that in more detail shortly.

I'll bet that most people don't have the space or wherewithal to install a full blown "L". I can well imagine there are some glum

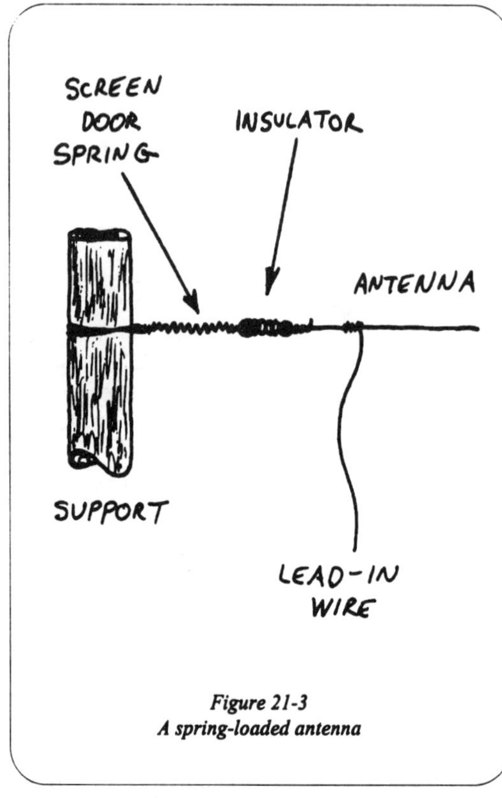

Figure 21-3
A spring-loaded antenna

apartment dwellers out there thinking, "Hundred foot wires? Twenty foot poles?!" Don't worry. The inverted "L" is a great antenna, but I don't have one, and never have. Lots of alternatives exist.

For instance, you can stretch a piece of wire from one corner of an attic to another. Use the same insulators and technique as you would for an outdoor "L". This approach is neat because the antenna is completely protected from the elements, and is invisible. (This may be a plus in dealing with a landlord, parents or a skeptical significant other.)

No attic? Try stretching a piece of bell wire from one corner of a room to the opposite corner. Anchor the ends of the wire as close to the ceiling as possible. Pins, staples, or thumbtacks make great anchors and won't tear up the plaster too badly.

In *The Giant Book Of Easy-To-Build Electronic Projects*, published by TAB, a radio enthusiast describes a really clever alternative to large, external antennas.

He was about to panel his den. Before the paneling went up, he taped a continuous strip of aluminum foil to the wall near the ceiling. He used the full width of foil, and unrolled enough to span three walls. He attached a wire to the foil, and paneled over the whole thing. I guess it worked just fine for his shortwave receiver.

All of the components in this book were tested with an antenna that consisted of nothing more than about fifteen feet of telephone wire, strung between the ceiling brackets that anchor my garage door. A heavy rubber band was knotted at each end, and the bands were slipped over the ends of the brackets. This kept the wire taught, and insulated it from the brackets, yet allowed it be compliant if something happened to strike it.

Even if you don't have the latitude to build *any* kind of antenna, you're still not entirely out of luck. Just about any large, ungrounded piece of metal will act as an antenna to some degree. Bed springs, a metal window frame, or an aluminum garden chair may work better than nothing at all.

In the days of dial phones, it was common knowledge that an alligator clip snapped on to the finger-stop on a standard phone dial made an excellent antenna for crystal work.

Years back, a fellow wrote about tinkering with an old radio when the telephone rang. He got up from his chair, but left his high impedance headphones on. He disconnected the leads, and let them dangle as he walked into the kitchen to answer the phone. Along the way, the headphone leads happened to touch a soup can on the counter. He was amazed to hear a burst of music.

To make a long story short, he claimed to be able to receive a certain local station on just about anything metal in the house! Presumably, oxides and minor corrosion on the metal surfaces acted as a detector.

Grounds are as vital to good crystal radio performance as antennas, yet they serve an entirely opposite purpose. While you might think of the antenna as a mechanism to collect energy, the ground is a mechanism to dissipate it. Now, why would you want to do that?

Consider the pinwheel example once more. At first glance you might say that the motion of the pinwheel is due to the wind that strikes the blades.

Think again. Wouldn't it be more accurate to say that the motion is really a product of the wind that passes *through* the blades? If so, then you could argue that the air *leaving* the blades is as important as the air going in, right?

Your antenna helps collect energy which passes through your radio circuitry. A good ground gives that energy somewhere to go. The work is done as the energy moves on it's way through your radio.

I'll concede to the experts in the audience that this is something of an oversimplification. One the other hand, it illustrates the relative importance of grounds without going overboard on details.

What makes a good ground? A classic ground is a long length

of copper rod, literally driven into the "ground". Electricians use copper-coated steel rods for the same purpose. Another approach is to bury large metal plates, panels, or mesh deep in the ground. The idea is to achieve a good electrical connection with the earth itself.

Obviously, a metal rod driven into moist earth will result in a better ground than an identical rod driven into parched earth. This has resulted in a rather interesting array of strategies using water, various salts and other chemicals, and even urine to ensure a good connection.

As was the case with antennas, there are alternatives to the classical grounds described above.

Cold water pipes can make excellent radio grounds. Why? First, metal conducts electricity, so in this case, the pipe may as well be a wire. Second, any cold water pipe can eventually be traced back to the well (deep in the earth) or a buried pipe delivering water from the city.

Some words of wisdom: If you choose to ground your radio to a water pipe, know what you are "hooking" up to. If your pipes are plastic, or if there is a section of plastic pipe between you and ground, this trick won't work. Also, don't use hot water pipes, and **NEVER** ground any equipment to a gas pipe!

In most houses, outlets and other fixtures are connected to earth ground by convention. This is done for safety purposes. A separate, bare wire stretches from every switch and outlet box back to the main fuse or breaker panel. From there, it's actually connected to a metal rod, driven into the earth. In theory, you can secure a radio ground by attaching a small wire to the screw that holds the cover plate on. Some people have reported good results using grounds of this type, others have complained of electrical noise.

I offer this information with some hesitation, in that, if you do something stupid, you could really hurt yourself. **NEVER** attempt to make any connection to the outlet, switch, or the wires themselves! Better yet, unless you are competent with and understand house wiring, forget I even mentioned it.

Sometimes, all you need for a ground is a large, conductive body. If those metal bedsprings don't work as an antenna, try them as a ground. For that matter, your own body works surprisingly well if you simply hold the ground wire in your hand.

Earlier, I mentioned the importance of using a lightning arrester with any outdoor antenna. Let me explain this just a bit.

To begin with, I don't know if *any* man-made contrivance can withstand a direct hit by a lightning bolt. Every year it sets buildings on fire, blasts trees to smithereens, and kills lots of animals and people. It's not something you *ever* want to mess with. When a storm rolls in, you'd best consider your radio fun **over**. Get away from your equipment, and resist the temptation to even touch it.

You don't have to experience a direct strike to suffer damage or injury. Lightning bolts are extremely powerful transmitters. A nearby strike can broadcast enough energy to make nearby antennas lethal to the touch. This is part of the reason for installing lightning arresters.

Lightning arresters are little black boxes that are connected between the antenna and a good outdoor ground. Under ordinary circumstances, they do nothing, and as far as your radio is concerned, they are invisible. However, if a significant electrical charge builds up on the antenna, the arrester will suddenly become conductive, and will direct that charge into the ground.

I've seen plans for a lightning arresters based on simple spark gaps. One design actually used an old spark plug. The ideas certainly seem feasible. On the other hand, where safety is concerned, I'd recommend going with a store-bought solution. Follow the instructions and documentation packaged with the arrester.

Page 168

Chapter 22
Sample Circuits And Assorted Ideas

By now I've introduced every basic component that you need to build yourself a working radio from nothing. You should now be able to capture a signal, tune it, detect it, and produce audible sounds with it.

I've suggested scores of ways to build these components, and outlined specific construction details for some of my favorites. The question at this point, is how to combine them into a receiver.

Experimentation is the key, and crystal radio components are wonderfully forgiving in this respect. You can connect them just about any way you please, without hurting anything. The radio will either work, or not work. You may hear several stations or none at all. You may hear signals other than those you expected.

None the less, in rounding out this book, I figured I'd toss out a few sample circuits to get you started.

Figure 20-1:
A Tuner-less Radio

This is about as elementary as it gets. The figure shows a basic long wire antenna, connected to a detector, headphones, and a ground. No coils are condensers are used to provide tuning.

This circuit works best in urban areas with nearby transmitters. Unfortunately, the absence of a tuner will probably mean that more than one station will come in at a time.

Even though there is no tuner circuit, it's possible that this radio

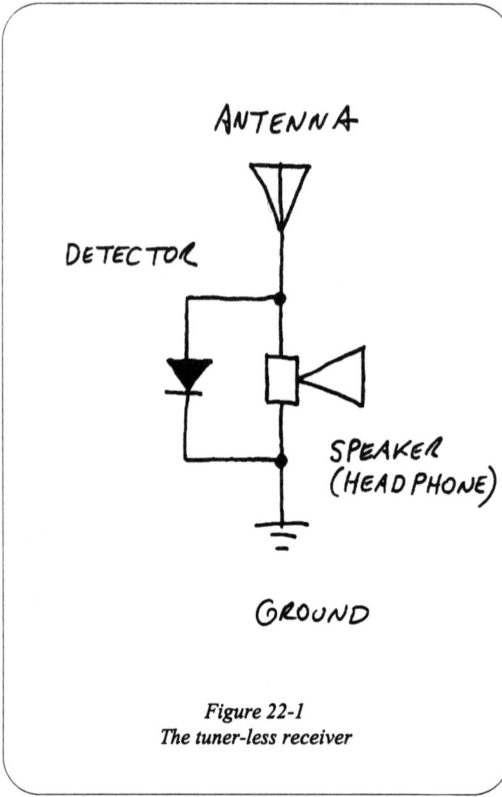

Figure 22-1
The tuner-less receiver

may favor certain stations. This is because of resonant factors in the antenna itself.

Figure 22-2 :
Using A Double Slider Coil

This figure shows a sample application for a slider coil like the Crank Coil. One slider is linked to a variable condenser to provide flexible tuning. The other slider helps compensate for the e l e c t r i c a l characteristics of your particular antenna.

Figure 22-3:
A Full Wave Detector

Simple detectors operate by rejecting one half of the received signal. One could argue that this is wasteful.   This circuit shows two detectors connected to a center-tapped coil. The idea is to try to drive the headphone with energy from signals of both polarities.

A set of this type requires more adjustment and tinkering, but the resulting volume in the headphones can be much louder.

Figure 22-4:
 Variable Bias on a Detector

Different detector materials have different thresholds of conduction. Some detector materials are exceptional in performance, yet require such a strong signal that they can't be used in practical radio sets.

One solution is provide an external voltage, called a bias. The bias is adjusted so that the detector is pushed very close, but not quite to the "on" state. Very weak radio signals can then drive the detector successfully.

Crystal sets that use detector biasing are unique in that a battery or some power source is required to operate the set. Fortunately, bias currents are very small, meaning that the power source can be just about anything. A common lemon, fitted with two different metal electrodes, may be a sufficient power source!

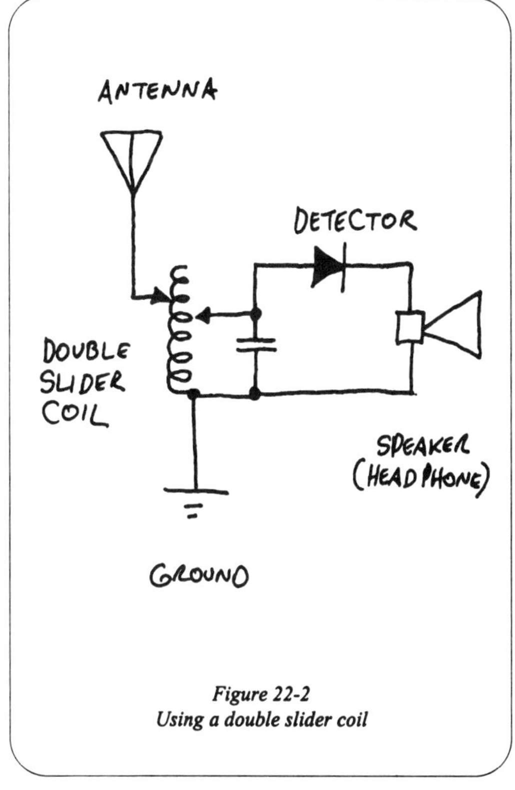

Figure 22-2
Using a double slider coil

The applied potential is made adjustable through the use of a variable resistor, also known as a *potentiometer*. Potentiometers are readily available if you want to play with this type of design.

Of course, using a store-bought potentiometer is "cheating" in the context of our original challenge. You must be wondering if I have a homemade version of this component, too.

Commercial potentiometers, or "pots" generally consist of a "C" shaped resistive element, and a rotating slider that can be moved to touch any point along the curve of the "C". The resistive component is generally made out of high-resistance wire, or a plastic material that is impregnated with carbon. The resistance between the two legs of the "C" is fixed, and depends on the dimensions and materials used in the resistive element. The resistance between either leg and the slider, however, is fully adjustable, and depends on the position of the slider.

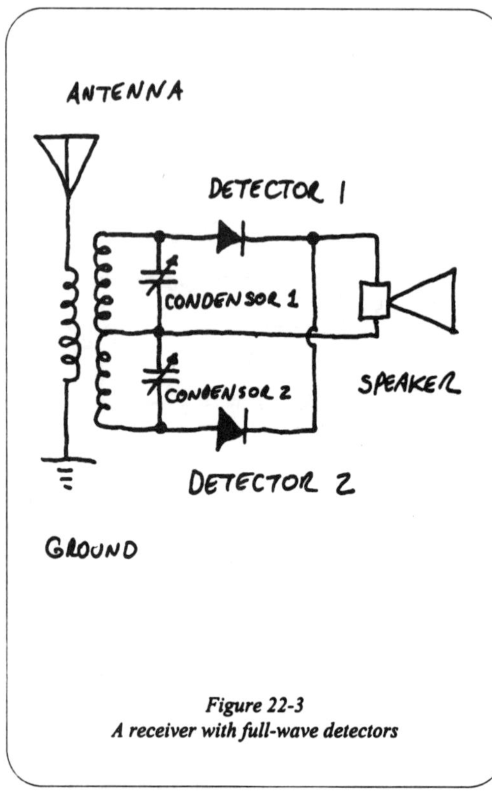

Figure 22-3
A receiver with full-wave detectors

I don't yet have a lab version of a pot, but I've tinkered a bit with what might be an ideal resistive element: a common "lead" pencil! (Pencils, of course, do not contain lead, but carbon and various binding agents.)

Start with a new pencil, and sharpen both ends to a point. The points will provide terminals, equivalent to the legs of the "C". How do you connect wires to pencil points? That's probably tough to do directly. On the other hand, you could glue the pencil down to a wooden base of some sort, and fashion two strips of springy brass, shaped to rest firmly against the pencil points. The brass strips would be terminated with binding posts.

Next, using a sharp knife, you need to whittle away the side of the pencil, down to the pencil lead itself. Use caution, first for your own safety, and second to prevent breaking or cracking the pencil lead anywhere along it's length. You also need to create a metal slider that will contact the exposed pencil lead, and slide up and down along it's length.

The pot in this circuit controls the bias voltage applied to the detector. The idea is to apply just enough voltage to bring the detector to the threshold of conduction, but no further. If you apply too much bias, the detector will turn fully on, and will no longer function as a detector. (Figure 22-4)

Another hint: If you have no luck with this circuit, try reversing the terminals of your detector.

Figure 22-5 shows a conceptual design for a pencil-based potentiometer.

The Beginning, Not The End

There you have it, as promised about twenty chapters ago: Fully functioning radio sets built from nothing more than scrap, and hardware store odds and ends.

The last page may signal the end of this book, but not the end of your education and fun. Read, and e x p e r i m e n t . Rediscover curiosity, creativity, and the fine art of thought.

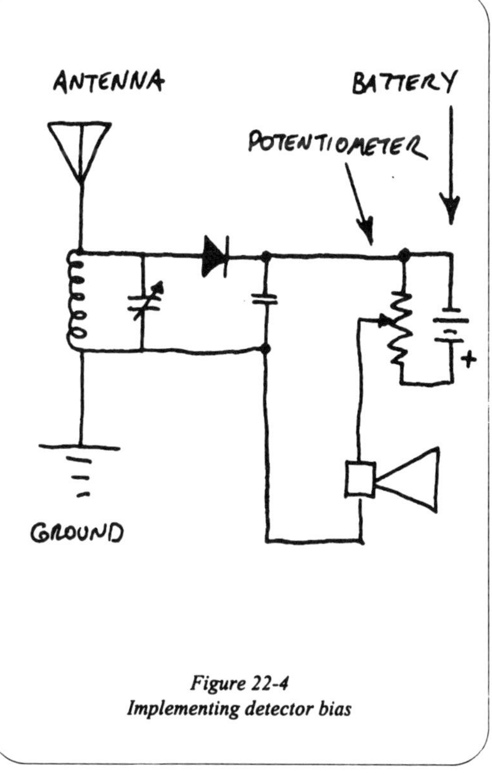

Figure 22-4
Implementing detector bias

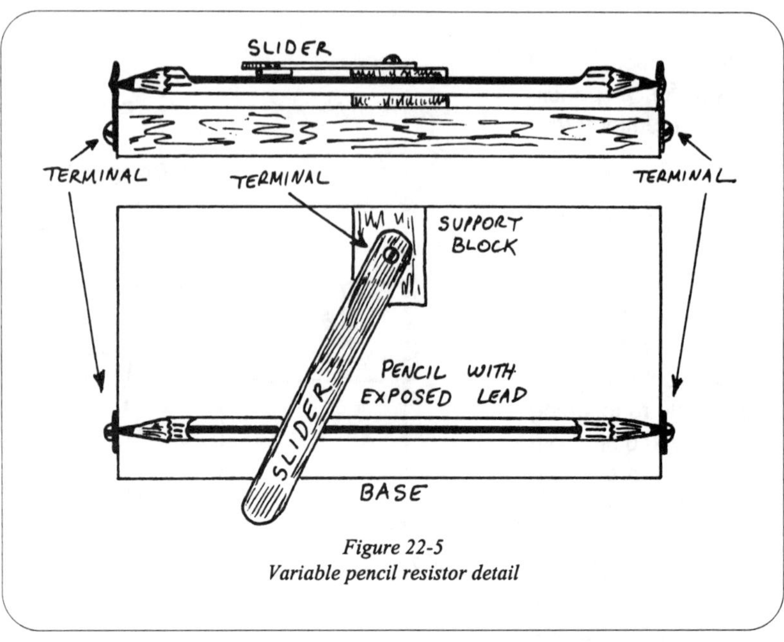

Figure 22-5
Variable pencil resistor detail

Appendix 1
Useful Formulas and Data

Permittivity	Permittivity
Vacuum	8.85×10^{12}
Air	8.85×10^{12}
Glass	$45\text{-}90 \times 10^{12}$
Rubber	$3\text{-}35 \times 10^{12}$
Mica	$27\text{-}54 \times 10^{12}$
Petroleum	18×10^{12}

$$C = \epsilon \frac{A}{d}$$

Formula for parallel plate condenser

C= capacitance in farads
epsilon=permittivity of dielectric
A= area of plates (meters)
d= distance between plates (meters)

$$L = \mu \frac{N^2 A}{l}$$

Formula for single-layer coil

L = Inductance in henries
mu = Permeability of core material
         (webers per amp-meter)
N = Number of turns of wire
A = Cross section area of core
l = Length of core

$$f = \frac{1}{2\pi\sqrt{LC}}$$

Formula for frequency

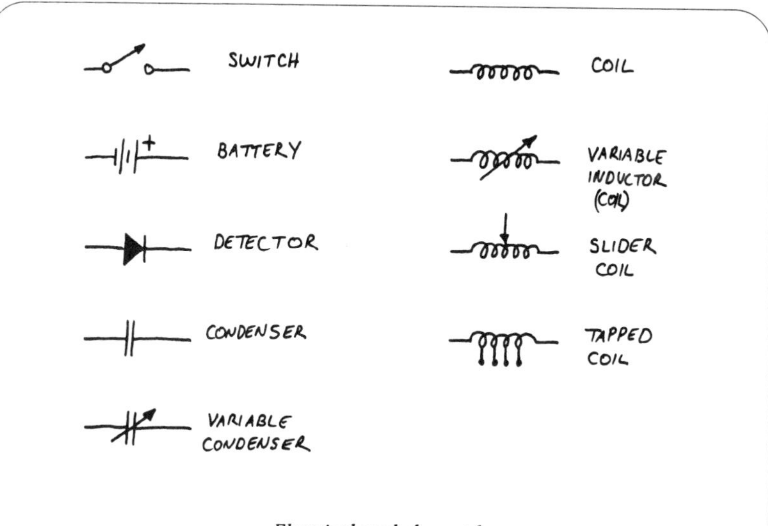

Electrical symbols part 1

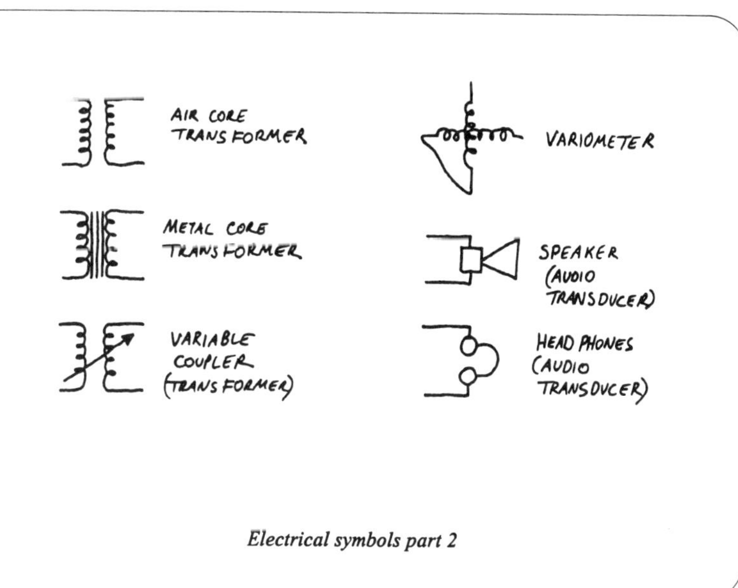

Electrical symbols part 2

Appendix 2
Where To Find Materials

At	You Can Find
Hobby/Craft Shop	brass sheet, brass tubing, small mechanical parts, wax, finished pine bases, brushes
Hardware Store	solder, aluminum and steel sheets, bolts, screws, threaded rod, and hardware. Stains, paints, adhesives, wire, plastic pipe, tools, magnets
Drug Store	piezo cigarette lighters, shoe laces, shoe polish cans
Supermarket	tuna and soup cans, foil, tape, wax.
Electronic Supply Store	Alligator clips, wire, solder, magnets
Fabric Store (Sewing supplies)	Felt, cloth
Office Supply Store	Tape, plastic sheeting, cardboard, adhesives, cardboard tubing, tape
Gas Station	piezo cigarette lighters
Garbage Can	Magnets, wire, metal, screws, bolts, nuts, hardware, glass, paper, metal, wood, tubing, electronic components
Guitar Shop	Old, cheap guitar pickups, for magnets and wire
Used Book Store	Excellent old books showing old equipment and circuitry

Appendix 3
References

The following are a small part of the excellent resources available in print and on the worldwide web (internet). Garage sales and used book stores are a great source of out-of-print electrical books that contain useful information.

Miscellaneous Web Resources

Website	Comments
http://www.antique-radio.org/class/class.html	Introduction to crystal radios with schematics
http://home.t-online.de/home/gollum/	German site, useful information, history, and schematics
http://connect.ccsn.edu/shs/build.htm	Construction details for a radio set
http://connix.com/~harry/drcrystl.htm	Crystal Receivers, Variocouplers and Variometers
http://www.ebay.com	On-line auction, frequently offers vintage parts and books
http://cpu.net/classicradio/radio_links.html	Radio collecting links
rec.radio.amateur.homebrew	ham radio newsgroup

Businesses

Business	Comments
The Xtal Set Society PO Box 3026 St. Louis, MO 63130 http://midnightscience.com/	Dedicated to experimentation with crystal radio technology. They publish a newsletter, and distribute books and materials.
Lindsay Publications Inc. PO Box 538 Bradley, IL 60915 815/935-5353 fax 815/935-5477 http://www.lindsaybks.com/	Outstanding resource for rare information, "lost" technologies, and reprints of outstanding vintage books.
Antique Electronic Supply 6221 S Maple Avenue Tempe, AZ 85283 Phone 602-820-5411 Fax 800-706-6789 (U.S. and Canada) or 602-820-4643 http://www.tubesandmore.com/	Vintage electronic parts. They stock many unusual and hard-to-find parts, including binding posts, high-impedance headphones, and various gauges of magnet wire.

Interesting Books

Title	Author	Publisher	Copy-right	Comments
100 Radio Hookups	Muhleman, Maurice	E.I. Co.	1924	Available in reprint from Lindsay Publications
Crystal Clear	Sievers, Maurice	Sonoran Publishing	1991	An amazing variety of documented vintage crystal radios
Crystals and Crystal Growing	Holden, Allan	The MIT Press	1997	Available in reprint from Lindsay Publications
The Crystal Set Handbook, Vol III	Anderson, Phillip	Xtal Set Society	1996	Available from Xtal Set Society
Crystal Set Projects		Xtal Set Society	1997	Available from Xtal Set Society
Hawkins Electrical Guide	Hawkins	Audel & Co	1927	I've seen several copies of this 10 volume set in used bookstores and on www.ebay.com
Henley's 222 Radio Circuit Designs	Anderson, John	Henley Publishing	1924	Available in reprint from Lindsay Publications
How To Make A Wireless Set	Windsor, H. H.	Popular Mechanics Co	1911	Very old designs for wireless telegraph equipment
Make It Yourself		Popular Mechanics Press	1927	Available in reprint from Lindsay Publications

Radio Engineering Handbook	Henney, Keith	McGraw-Hill	1941	Outstanding resource work
Radio Receiver Projects You Can Build	Davidson, Homer	TAB Books	1993	Also contains plans for tube and transistor radios
Radio Simplified	Kendall, Lewis F.	John Winston Company	1923	Primarily concerns crystal and simple 1 tube radios
Radios That Work For Free	Edwards, K.E.	Hope & Allen Publishing	1977